TELECOMMUNICATIONS, VALUES, AND THE PUBLIC INTEREST

COMMUNICATION AND INFORMATION SCIENCE

Edited by
Brenda Dervin
The Ohio State University

Recent Titles

TELECOMMUNICATIONS, VALUES, AND THE PUBLIC INTEREST

Edited by

Sven B. Lundstedt

ABLEX PUBLISHING CORPORATION
NORWOOD, NEW JERSEY

Copyright © 1990 by Ablex Publishing Corporation

Library of Congress Cataloging-in-Publication Data

Telecommunications, values, and the public interest / editor, Sven B.
  Lundstedt.
        p.   cm. — (Communication and information science)
    Based on papers from a roundtable held in 1987–88 under the
  auspices of the Graduate School at the Ohio State University.
    Includes bibliographical references and index.
    ISBN 0–89391–693–5 (cloth); 0–89391–733–8 (paper)
    1. Telecommunication—Social aspects—Congresses.   2. Public
  interest—Congresses.   3. Values—Congresses.   I. Lundstedt, Sven
  B., 1926–.   II. Ohio State University. Graduate School.
  III. Series.
  HE7604.T458   1990
  302.2—dc20   90–999
                                                                  CIP

Ablex Publishing Corporation
355 Chestnut Street
Norwood, New Jersey 07648

Table of Contents

Foreword

The "information society" is upon us. The development and use of telecommunications technology have literally exploded over the past 10 to 15 years and accelerated our dependence upon ready access to and use of information. The telecommunications industry, broadly defined, makes up over four percent of our GNP. Telecommunications affects the lives of us all, with or without our awareness and approval.

A roundtable organized by Professor Sven B. Lundstedt, "Telecommunications, Values and the Public Interest", was indeed a timely response to what some see as an unaddressed question in telecommunications. Is telecommunications technology forcing a fundamental redefinition of values in our society? Issues of privacy, access, control, and internationalization take on a new meaning in the face of the rapid technological developments in telecommunications.

Professor Lundstedt has assembled an outstanding group of scholars and practitioners to bring their perspectives to bear on these issues. The roundtable in which they participated was held at The Ohio State University during the Winter Quarter of 1988, supported by a grant to Professor Lundstedt from the Ohio State University Ameritech Fellowship Program. The roundtable brought together scholars from around the campus and the nation who, in many cases, had not known one another previously. One outcome was the development of several networks of scholars whose interaction and collaboration have continued long after the conclusion of the roundtable. This book is another tangible outcome, containing chapters based on presentations during the roundtable.

Clearly, technological developments in telecommunications have presented realistic options previously not even imagined. For example, commercial transactions using telecommunications technology, such as banking and shopping without leaving home and in some cases without personal interaction, are available to most people in the country. Transfer of large amounts of information across long distances and international boundaries is commonplace. Electronic access to personal and corporate records reduces the need for paper transactions and clerical intervention. Management of information and public opinion in political campaigns has taken on new dimensions.

Each new option, however, has attendant complications. What are the potential economic, social, and psychological impacts of these developments? How is the nature of economic and political exchange affected? How is national security compromised by ready international access to data banks? What are the risks to individual privacy? Are answers to these questions different in developing nations? Many more examples could be cited. But in all cases, there are unanticipated challenges to our value systems and conflicts among values, all of which have champions.

This volume addresses some value implications of telecommunications technology development. The authors will challenge scholars and policy makers alike who wrestle with the changes brought on by the telecommunications revolution. Professor Lundstedt and his collaborators have provided a significant stimulus to our thinking about the impact of telecommunications on values and policy making in our society.

Roy A. Koenigsknecht
Dean of the Graduate School
The Ohio State University

Preface

This book is based upon a Roundtable on Telecommunications, Values, and The Public Interest held during the academic year 1987–1988 at The Ohio State University. With a few exceptions these chapters were written by members of that roundtable. We are grateful to all the members of the roundtable for their participation and contributions. Funding for the roundtable activity came from the Graduate School at The Ohio State University as a result of a generous grant from the Ameritech Foundation in Chicago, Illinois.

One of the most important issues in modern telecommunications is its relationship, as a preeminently social technology, to values and to the public interest. The inexorable march of technological innovations in communication, driven by powerful economic forces, has had a tendency to sweep away all else before it. We have already seen some of the unwanted consequences of this enormous force in the modern world in the form of second and third order consequences illustrated in the media orientation of recent presidential elections in the United States, a critical example of the link between values and the public interest. But these consequences are much more far-reaching than even these important events, as the chapters in this book illustrate. So there cannot be any disagreement about the essential importance of this subject. In another sense, this volume is an exploration of these critical issues in entirely new ways, and a thoughtful reflection on the forward movement of this form of technology. Perhaps also these chapters should raise serious questions about this technology's direction and eventual impact, both good and bad.

I wish to thank the Dean of the Graduate School at Ohio State, Roy Koenigsknecht, who has prepared the foreword to this book, for his continuing assistance and thoughtful vision and leadership in the Ameritech program at Ohio State and Paul Isaac, Associate Dean in the Graduate School, for his skillful administrative guidance of the Ameritech Program. To my colleagues on the Ameritech Advisory Committee, Professors Douglas Jones, Thomas McCain, and George Smith, and Associate Vice President Thomas Sweeney, go thanks for their support, encouragement, and occasional useful criticism. I wish also to thank Associate Dean Astrid Merget, Director of the School of Public Policy and Management, for her flexibility and willingness to permit me to enjoy the necessary released time to complete this project. And I wish to thank Professor Brenda Dervin, who, as the Chairperson of the Department of Communication, provided intellectual support and encouragement. To Hee Soo Kim go my thanks for his assistance with the administrative details of the Telecommunications Roundtable. And last, but certainly not least, I want to thank Edward Jennings, President of The Ohio State University, who has sought to create an overall university environment that enhances the free exchange of ideas no matter what their intellectual origin may be or seem to be, and no matter how controversial.

Sven B. Lundstedt
Columbus, Ohio
December, 1989

chapter 1
Introduction

Sven B. Lundstedt
Ameritech Research Professor
Professor of Public Policy and Management, and International Business
School of Public Policy and Management and Faculty of Human
  Resources and Management
The Ohio State University

The purpose of this book is to explore some of the important connections between telecommunications, information technology, human values, and public policy. It assumes that values and technology are interrelated, and that the invention, design, manufacturing, and uses of any technology can never be completely value free.

A value can be defined as an element of a person's system of beliefs, feelings, and tendencies to act toward people, things, ideas, and concepts and any other facet of the natural world. There are as many values as there are people and object relationships in the world, and as there are institutions, organizations, and small groups which serve as repositories of them. Values are also part of coherent systems of beliefs held by people and as reflected in their culture. The opposite may also be true to the extent that incoherent value systems may develop during periods of transition and social change when value systems, for example, may temporarily seem disjointed and confused. Because values may be deeply habitual aspects of human behavior, people may be unaware and unconscious of them. People may also be exceedingly conscious of them where a strongly held ideology exists.

Much has been written about human values through the centuries. The literature in the social and behavioral sciences and humanities, including philosophy, theology, and religion, is extensive. Research about values is especially common in psychology, anthropology and sociology. They are studied at length in economics and political science. For centuries, philosophy, including aesthetics and theology, have been concerned with values, as well as with history, literature, and the performing arts. I will not attempt to review this vast field in this brief introduction. For a subject about which so much has been written for so long, the absence of any reference to their critical role in many other fields is surprising.

Organized in five sections, this book first looks at some aspects of the overall value environment of telecommunications and information technology, and then economic and business perspectives, space technology, economic development, and, finally, normative considerations.

THE VALUE ENVIRONMENT

Harvey Brooks begins this exploration in Chapter 2 by examining unrecognized consequences of telecommunications technology. He approaches his subject from the point of view of the broader field of information technology rather than from the narrower sense of telecommunication services for discrete subscribers. One of his conclusions is that the most significant development of the last decade has been the rapid change from a technocratically planned telecommunications infrastructure to one governed increasingly by market forces. But, there is uncertainty concerning the distributional significance of this trend and whether it will be successful in meeting communication needs equitably. The economic values associated with a purely market approach are questioned, especially the assumption that social benefits can be equated simply with the sum of private benefits.

Brooks asks if "information power" may become more concentrated and less accessible as a result of technological change under exclusively market forces. He raises important issues concerning governance and equity when he suggests that information technologies (IT) and telecommunications may broadly reenforce the present social class structure of the United States, leading to further inequality. In his consideration of organizational centralization versus decentralization, he discusses optimal organizational structures for efficient and effective productivity management.

In the production sectors of the economy, decentralization may be more functional than centralization because "responsibility is better matched to detailed operational knowledge." This important point recognizes the changes in instrumental values underlying management and organizational behavior. Flexible specialization and other new work patterns serve as examples of the increasing diversity in the structure of human work.

The role of information technology in the global dispersion of manufacturing may, however, provide only limited advantages in the economies of developing countries. Given the formidable problems of local manufacturing logistics, such as transportation and quality control, it is not clear if the cost advantages of cheap labor alone will contribute to greater efficiencies of geographical agglomeration of production complexes.

Information technology also has had significant effects on governance. One of them is the possible limit on public decision making created by rapid communication, as well as the need to shorten messages by the media to make them fit both economic and audience requirements. Recent presidential debates illustrate how political communication has become less spontaneous because candidates are now responding to vast unseen audiences much as one might in advertising or entertainment. Have the traditional values which encourage substantive open and candid political debate of issues weakened as a result of candidates playing to mass television audiences? Information overload is also a serious problem affecting public decision making.

In Chapter 3 Chadwick Alger examines the relationship between telecommunications, self-determination, and world peace. Self-determination is the extensive striving for nationalistic self-determination. Since World War II, self-determination has increased the demand for information resulting in a diffusion of new political values.

In this diffusion an important value issue is information equity. The free flow of information between nations underlies the larger problem of how to create a functional instrumental value system that all parties can be guided by in negotiations. This has been a major technical and political goal in the United Nations. His chapter provides illustrations of how information, transmitted in international conversations, plays an increasingly significant educational role to bring about those value changes seen as important in modernization.

A benefit of telecommunications, in particular, and information technologies, generally, can be to reduce communication barriers to mutual understanding and problem solving between nations arising from differences in cultural patterns and values orientations. Improved communication will not only influence formation and implementation of national goals toward peace and modernization, but will help to increase greater mutual understanding among different cultures and nations.

The presence of instrumental values (those which are, for example, related to human behavior such as cooperation) which support economic, social and artistic cooperation, in contrast to military conflict, are important to maintaining a sustainable peace. To the extent that new communication technologies can help to resolve value conflicts through education, they may also create other opportunities for peaceful relations through mutual understanding. Shared instrumental values about the moderation and control of conflict may result from an improved learning environment assisted by telecommunications. The examples in this chapter illustrate the role of telecommunications in advancing self-determination and world peace.

The importance of language in international communication is described in Chapter 4 by Johanna De Stefano. She concludes that the English language plays a critical role in a process of socialization and formation of values that influence sovereignty and economic development. It plays an important, even dominating, role in many less developed countries.

Does the English language serve as a "mirror" reflecting British and American values? Interesting questions are raised about the role of telecommunications in spreading culture, and political and economic influence, through use of a dominant language. Does English language act as a *window*, that is, as a transparent neutral aspect of communication and a ubiquitous tool for anyone's use, or does it serve as a "kaleidoscope" through which a message is distorted conveying different values opaquely communicated but not always clearly understood?

These are questions of basic importance about human communications that have implications for the ethical use of telecommunications technology. Where will this lead Third World countries who seek to formulate new economic and social values to develop their sovereignty and programs for modernization? Which languages will dominate commercial and diplomatic relations is of considerable practical importance.

In a fifth chapter, written by the book's editor, information technology, privacy, and the public interest are explored in the light of values underlying the Constitution and American political system. This discussion attempts to show how the instrumental values of a democratic society are used to protect ultimate political values from the abuses of a wrongly used information technology. The phenomenon of value *erosion* or *decay* is also possible in the conflict between basic normative values of privacy, upheld by the general values of the Constitution, and the commercially expedient values of the marketplace, which suggests that privacy can be for sale. Frequent reaffirmations of the original written sources of the ultimate values by the higher courts, the Congress, and the President are very important. Ordinary patriotism is also such a reaffirmation.

The final chapter in this first section, by Joseph Pelton, concludes that something called *telepower*, economic and technological in its form, is an important underlying factor in the growth of telecommunications and information technology, and in their impact upon society. Pelton addresses the question of telepower in terms of information and services, as a facet of economic development broadly defined, and telecommunications.

In a summary of public interest issues, Pelton suggests legislative reforms are needed about such value-laden issues as technological unemployment, off-shore electronic services, international financial trading, abuses of electronic monitoring, trends and development in electronic crime, and the constructive use of personal computers and other electronic devices to monitor privacy, and control of information processing of wanted and unwanted information. Supporting the previous discussion in Chapter 5 on privacy, he mentions the unauthorized access to personal computer files and lack of access to one's own computer files as undesirable consequences. In these examples of the value environment of telecommunications and information technology, we see value issues that are encompassed by an enormously complex social environment, far reaching in significance and affecting fundamental human institutions and culture.

CORPORATIONS, BUSINESS, AND THE COMMERCIAL
INFORMATION REVOLUTION

In Chapter 7, Joseph Daleiden reflects upon some of the social value issues important in the development of corporate-government telecommunication policies. Starting from the premise that people seek to increase their happiness or to minimize their unhappiness, he argues that specific actions which lead to happiness can be evaluated in terms of the degree to which they contribute to that goal. But conflicts will always arise, usually over questions of relative value choices. The purpose of an enlightened social policy intended to serve the public interest is, in part, to help resolve such conflicts in the arena of economic life. Rawl's theory of justice provides a starting point for developing and evaluating the value bases of such social policies.

Daleiden says that commercial competition is both a function of the market and the policies that protect the market. Turning to recent telecommunications policy, he concludes that it has been too concerned with competition as an end in itself and not enough with an assessment of the differential impacts of such policies on the welfare of various interest groups in society. He maintains that in order to maximize in synergistic fashion the benefits of new technology, while protecting the goal of affordable, universal, telephone services, the market must be constrained in some measure by a social contract.

Speaking of the regulatory system of the Federal government, this would supposedly mean a social contract expressing certain political and economic values. He argues that the Bell operating companies would be allowed to enter any market and offer any services so long as they offered universal telephone services, too, within price ceilings set by regulatory bodies.

In Chapter 8, addressing value issues in transborder telecommunications, Riad Ajami says that transborder data flows are one of the most critical issues facing international trade in services. Important driving forces include pressures arising from the shift to a service economy, the internalization of markets, and growth in use of computer processing in telecommunications. For economic efficiency to take place, the transfer of information across national borders should be free of restrictive regulations.

Ajami reviews the different kinds of transborder data flows and then examines economic and noneconomic value issues of concern. He concludes that the two principle forces of politics and economics should be subjected to comparative value analysis, presumably, I would add, to avoid serious value conflicts over instrumental and goal values. The values of autonomy, pluralism, and diversity guide national states, while economic imperatives of effectiveness, efficiency, and timeliness guide the strategic value alternatives for multinational corporations. The two value systems often do not line up with one another.

His chapter provides another kind of global perspective that emphasizes the difficulty of describing rapidly changing economic systems. Without broader cultural value perspectives, it will be difficult indeed to understand the value problems, and the less visible forces, that underlie economic and social changes.

I can add another important current example. An increase in transborder flows is a key aspect of the European Economic Community's [EEC] efforts to have open borders by 1992. Coming from the EEC deliberations in Brussels are a variety of instrumental policies aimed at reaching this goal. Along with this wish list is an optimistic forecast of future economic growth. But, deep-seated values underlying national sovereignty still are present and they may successfully block progress.

Deeply rooted cultural and political values developed over centuries of European conflict and cooperation to preserve or expand national sovereignty and economic power may be stronger than the optimistic forecast provided by EEC public relations in the long run. History records that there have been many other regionalizing efforts in the past, albeit military in nature in most cases, but also commercial as in the case of the Hanseatic League in the 13th century. All declined eventually, in part because of value incursions from without and value changes from within, many of which were unforeseen at the time.

In Chapter 9, W. Wayne Talarzyk and Robert E. Widing II discuss values which shape the role of telecommunications in retail by use of videotex. The importance of what they say can be appreciated by noting the rapid changes in the way in which consumer needs are being satisfied and enhanced by telecommunications. They show how telecommunications can reduce time, space, and merchandising barriers between retailers and consumers. The economic implications and side effects include a reduction in prices as well as a better use of time and effort. They discuss numerous ways in which telecommunications-based technologies improve retail operations and thereby lead to improved economic efficiencies.

They also examine telecommunication methods that involve marketing to consumers, followed by a discussion of technology-driven innovations reflecting a search for consumer values that can be satisfied by joining them with existing and emerging technologies. The chapter closes with a discussion of public policy issues related to use of telecommunications in marketing particularly with reference to videotex.

Chapter 10 in this section, by Marilyn Greenwald, questions whether the videotex market has future economic potential. She discusses the power of values to shape attitudes toward product services adoption, and concludes that the simple efficiency of a new technology may not be enough to ensure its eventual success. Resistance to change creates a reluctance to alter behavior and so has hindered growth of videotex. The pattern of resistance seems to be interactive, with a responsive negativity shared by some private companies, public agencies, and media because they anticipate a threat to their investment in a new product, or the vested interest in older ones.

This chapter, and others as well, illustrates the importance of understanding the social and psychological forces which determine the underlying values which shape the life cycles of commercialized technologies.

I would add that videotex, as such, is a technological innovation which requires social innovations to be successful, just as the automobile required social inventions in the form of service networks to be used successfully. The psychological changes will usually require a readiness to learn new behavior and attitudes concerning new technologies. As such, learning itself also has to be seen as a valued process.

Since this has more or less been true of all technological changes throughout history, what seems to be required in technology adoption is to arrive at a point of development of a technology where the instrumental and the goal values concerning what may not yet be perceived widely as being rewarding and useful to people can become quickly learned, and perceived, to be useful and rewarding. Such a change requires both instrumental value changes (consumer education through advertising), as well as changes in other value modalities that may be aesthetic and utilitarian, as well as not be in dissonance with other generally accepted values at the time. Fashion does play a role, and the fashion industry, for example, must surely be one of the most value conscious of all industries.

TELECOMMUNICATIONS AND SPACE

Value issues, particularly ones coming in the latter third of the 20th century, are discussed in Chapters 11 and 12. They include telecommunication technology in space and the use of operations research (OR) models for synthesizing communication satellites. Molly Macauley examines from an economic perspective the use of outer space for acquiring and transmitting information in the global information economy. Charles Reilly, writing about operations research models (OR) for satellite synthesis from the perspective of industrial engineering, discusses five values associated with operations research: equitableness, conservation, efficiency, flexibility, and practicality. OR models are considered to be complimentary methods helpful in negotiations about satellite communication policies.

Macauley's selection of value categories, which she says are inherent in public policy development for space and terrestrial communications, include awareness, convenience, autonomy, fairness, wealth distribution, prestige, leadership, resource use, efficiency, standard of living, privacy, environmental quality, culture, and technological progress. These modalities seem to be fairly general ones, even fitting all societies, but surely take on different specific forms in each.

For example, some less-developed countries [LDCs] proclaim a valued right to have their own satellite in space, rather than to have to rent that space from consortia based in the technologically more advanced societies. But the cultural values of many LDCs are in fact not in tune with other instrumental values of technological efficiency necessary for the support and use of advanced satellite telecommunications. One policy limitation is that although their political and economic hopes for rapid advancement may be understandable, the LDCs have yet to demonstrate an ability to use even an existing basic telecommunication technology such as the ordinary telephone. So storing up space in space probably serves mainly as a political act rather than one of economic and technological efficiency in most cases.

A common error is that value neutrality, the view that assumes that anything, including scientific and technological methods and discoveries, can be viewed in a value-free way, is still often assumed to be possible. In fact, from a subjective or objective psychological perspective, value neutrality about anything is probably impossible to achieve since we are always the captives of our own values.

However if on the other hand, value neutrality means to be aware of, and tolerant toward, the values of others, regardless of how different their values may seem, and to be able to suspend critical judgment about these values, then a form of comparative value neutrality is possible, and even desirable, where the goal may be to reduce conflict with others and increase understanding. Perhaps the idea of value tolerance and comparative value awareness is a more accurate and appropriate way to think about this issue when one is trying to understand someone else's values, rather than judge them only in terms of one's own.

Consequently, the view that technology, or pure science for that matter, has little to do with values is an absurd assumption to try to prove. To do so may only blind one to the potential good or evil of technology's side effects. It is also naive to be unaware of the very important encoding functions of values in human perception and judgment, and in the ultimate adaptation and survival of the organism.

Errors of omission and commission that have been made while assuming that scientific and technological innovations are somehow value free, have led to severe damage to the global environment. The present concept of technology assessment serves as an excellent example of a form of value assessment of policy.

ECONOMIC DEVELOPMENT

Chapter 13, 14, and 15 in this section illustrate several value issues concerned with the development of telecommunications and information technology in other societies, as well as in our own. In his chapter on telecommunication technologies and social change, Tetsunori Koizumi discusses Japanese fascination with Western communication technologies in terms of a typology of evolutionary change.

He asks: What is the nature of social change that recent advances in telecommunications technology have brought about in Japan? In his analysis he uses four constructs of evolutionary change: growth and decay, structural change, synergistic transformation, and transmutation. He argues that telecommunication is an aspect of a form of *third wave* and has contributed to two kinds of social and economic change—synergistic transformation and transmutation. All four have counterparts in human values.

He draws attention to the importance in economic development of a wider framework in terms of which to understand the underlying mechanism of Japanese social processes which govern, and are governed by, new technology such as telecommunication. Against this larger background the transmutation processes that have led from the traditional telephone industry to the global information society can be better understood.

His discussion of the Japanese social system is especially insightful. He concludes that Japan is in a second phase of a synergistic transformation, or *informationalization* of its society, as the Japanese say. A single event, the synergistic transformation of the relationship between government and business symbolized by the 1984 Japanese telecommunication business law, prepared the way for the privatization of the Nippon Telegraph and Telephone Company (NTT). As telephones and computers are integrated, NTT has started a second life as a private corporation with NTT serving also as an agent of social and economic change in creating the Japanese "information network society." The accompanying value changes in Japan are judged to be very significant.

A different perspective is offered by Rohan Samarajiva and Peter Shields in Chapter 14 on value issues in telecommunication resource allocation in the Third World. Telecommunication resource allocation decisions may be seen in terms of supply decisions between investing in telecommunication as against other things; as demand decisions where choices are made between classes of users; and as another level of choice involving cost and price relationships. These decisions, be they made by government or by private firms, involve choices that affect others in society, and are hence considered to be value laden. The authors pay particular attention to the writings of policy experts who advise and inform decision makers. This chapter seeks to lay bare key value issues pertaining to academic and policy writings on telecommunication policy in the Third World.

Samarajiva and Shields understand values as normative judgments of good and bad. They concede the absence of objective criteria for value judgments and, therefore, the inability of all sectors of society to agree on all values. Their position is that there is a need for open discussion of values underlying resource allocation decisions, whatever those values may be. They base this conclusion on the value judgment that participation is the proper instrumental means to reach value consensus sufficient in allowing negotiation. Referring to the perceived difficulties of assuring broad participation in complex technological and investment decisions, such as telecommunication, they argue that constructive popular participation can be 'assured by making value issues explicit, as they have attempted to do, and by opening them up for discussion. They contend that the laying out of value implications, and the provision of opportunities for open and informed discussion, must be institutionalized as part of the planning process.

W. Richard Goe and Martin Kenney discuss the effects that information technology has on the United States political economy. They discuss specific applications of information technology and how they are transforming the ways in which information is used to promote the manufacture and distribution of goods and services. This chapter has been included in this section instead of in the section about business and corporations (it could presumably fit in either one) because it offers a salient analysis of economic development in the United States, thereby providing a clearer profile of such growth in an advanced Western economy.

Although direct comparisons with the LDCs are not made, one can imagine the profound differences and difficulties in economic development that lay ahead for the LDCs, many of whom want to acquire Western technology and economic values without fully realizing that by doing so they are exposed to value erosion and a substitution of many of their own cherished values. Rather than being old and useless these values may be quite excellent and have a high intrinsic quality of life value for them.

The so-called privatization movement is another major economic and political value issue that has reached international proportions. To what extent can it be said to be a significant economic force in the LDCs is an interesting question. The sharp ideological value conflict posed by recent privatization trends is reminiscent of value conflicts between command and market economies over the last 50 years. Perhaps in this value conflict there are no real winners or losers because everyone may be just part of a larger value transformation and synergism now seeking a new balance and center of gravity, of which the dramatic changes in Eastern Europe are but one example.

The command economies have discovered the economic weakness and unproductivity of too much ingrown bureaucracy and political control. At the other extreme, those libertarian advocates of a completely free market may find it necessary to come to terms with the fact that a totally market-oriented economy is an impossibility, even an anacronism. The example of the recent failures of the Soviet economy will send signals that doctrinaire economic orthodoxy is no longer a value orientation that has as much economic utility as does political and economic pragmatism, a form of utilitarian thinking. Even though some followers of the Reagan administration's economic policies have in the past decade vociferously advocated a free market, the United States has never really been an orthodox free-market economy, notwithstanding the strident views of some economic extremists who advocate the purest possible market economy. Hence, the United States economy is quite accurately and realistically often described as a mixed economy.

The role of telecommunications in economic and social change is significant not only insofar as it is an engine of that change, but also because, through the media of television and radio, the drama of change is being played out at a symbolic level. The media is a vast learning environment (I would even go so far as to call it a global electronic school) whose communicated values taken as a whole must be considered to be one of the most potent psychological forces in this century. If the changes in this half of the 20th century are part of a "long wave" of socioeconomic change, what will the future synthesis be like? And what role will telecommunications as a disseminator and "teacher" of values play?

NORMATIVE VALUE ISSUES

The final section of this book, Chapters 16 and 17, is concerned with the other side of the privatization issue, namely, control through regulation of its economic side effects. These chapters do not pretend to be a definitive discussion of what is a vast and complex subject of literature. The legal side of regulation alone occupies an enormous portion of it. Consequently, these final chapters focusing on the Federal Communications Commission (FCC) in the United States are intended only to illustrate certain unique value problems.

In Chapter 16, Joseph Foley discusses how the FCC has created a lively marketplace for the transfer of broadcast station licenses. He examines the values which are part of these FCC policies. In this case, the balancing of the rights of broadcasters and the public presents a significant value conflict. In one example, Foley refers to the sale of broadcast license transfers and raises the question—to whom should the revenue from them be given, the private firm or the government? It is essentially another varient of the old value conflict which reflects disagreement over whether government regulation or market forces should dominate and limit use of the broadcast spectrum.

If indeed, the use of the broadcast spectrum is to share in a public good versus a private one, then the value issues raised in this chapter are among the most critical from the point of view of national regulatory policy. The reference to price as one of the key determinants of market policy raises important questions concerning the possible ignorance of, or dilatory attitudes toward, a deeper sense of values in American society.

Price as a determinant of overall value may also drive out and limit the quality of media broadcast programming, because it is my impression that the television market is guided by a population response average closely associated with statistical distributions of consumer demand. This would appear to be manifested also in a sort of statistical regression toward the consumer market population mean, or least common denominator of values. Those who might argue against this form of criticism in the media could cite examples to the contrary, perhaps acting as apologists by dwelling more on what seems to happen at the low incidence tails of the population distribution then on its plump and profitable, but lower quality population average. Advertising income helps to make it all happen, so they might say.

However, it is visibly true that the popular "bread and circuses" approach to the broadcast market can have a depressing effect upon broadcasting program quality if that level of quality, by contrast, leaves out too much programming that is inspirationally educational and cultural in nature, but low in popularity ratings. The media is a very tough business and highly competitive. It is easy to see how standards can slip under economic pressures. And it is easy to see how the rationalization that television only gives people what they want can be seductive.

We may puzzle over the value dilemmas illustrated by television's outstanding commercial success. What precisely is the wider public interest in this case? How do we begin to define it? And what are the normative bases for judging whether it is or is not being met? These are issues that might more easily come up in Great Britain in the shadow of the BBC than in the United States.

It is a cultural imperative of the United States, and many other countries, that the public interest is a central value determinant in the all-important process of governance. To the extent that important basic value considerations of how a public good should be allocated are not included in the value agenda, and left unexamined and obscured by a narrowly defined market criteria such as price or profit, we can expect the value problems of governance also to arise. But to offset the perception that this is merely another outburst of criticism, I am reminded by those many elevating times when highest programming standards are consistently nourishing the public interest on television. Texaco's 50-year support of the Metropolitan opera is a case in point.

Another important consideration is whether the underlying policy values to be debated in the so-called open forums of our society are indeed visible enough to be recognized. We may need concepts and terms which distinguish between policies that clearly are visible and those that are not. Open communication and the free flow of information that is transparent are important values in American culture. In the final chapter, by Lundstedt and Spicer, the issue of latent policy, as contrasted with manifest policy (mentioned earlier in Chapter 5), and the FCC is discussed.

I have defined *latent policy*, in contrast to *manifest policy*, to mean a policy that is intentionally or unintentionally concealed from others. The two kinds of policy can further be distinguished by whether they are *divergent* or *convergent* when compared with a focal policy. A policy can further be identified according to whether it is *benevolent* or *malevolent*. Manifest convergence or divergence represents agreement and disagreement in open communication. This reflects values concerning the free and open exchange of information. Latent convergence is a form of tacit agreement with a focal policy and values, and could be either benevolent or malevolent.

But, latent divergence (secret disagreement), while not always resulting from malevolent intentions, can also arise as a consequence of fear and mistrust, misunderstanding, poor communication, policy and value conflicts, ulterior motives and covert secret activities. I would think that latent divergence from a focal policy, or secrecy in the act of disagreement and opposition to a policy, is potentially a very dangerous tendency in a democracy where it may supplant open communication, and may lead to covert activities which seek to subvert.

In Western democracies, and increasingly in some of the more authoritarian societies, the efficiency of the political and economic system is increasingly dependent upon accurate information. Moreover, ordinary business with others in an increasingly global economy cannot be transacted efficiently without policies that encourage trust and dependability in communications. Neither can complicated diplomatic and political transactions and communications be conducted efficiently as a rule.

But, to some extent, secrecy and other forms of privacy, taking the form of latent divergent and convergent communication and policy, deserves protection too and is a necessary and highly valued part of human communication. Closed-door sessions of any legislative body, well-intentioned private conversation between two people, and client-professional relations, for example, should be protected for a variety of very good reasons.

As in the past, responsible members of society need to be aware of whether the system, or individuals in it, are reaching dangerously seditious, or otherwise harmful, limits of latent divergence; or whether it is a truly malevolent form of communication and policy formation and implementation, as the Watergate and Iran-Contra scandals illustrate. Clarity about basic political value is essential in determining when that limit has been reached. Telecommunications that reflect a high degree of secrecy have a potential of being dangerous, particularly if important communications occur about public matters of concern without the knowledge of listeners and viewers.

The final chapter about latent policy forms a closing point for this discussion of telecommunications, information technology, values, and the public interest. The book as a whole draws attention not only to the dangers associated with the enormous power of telecommunications in government and business bureaucracies, but to the positive values and discoveries associated with telecommunications and information technology; especially some of the ways in which it has improved the quality of life in this century, and will continue to do so in the next. There is also an important need for continuing research.

The Value Environment

chapter 2
Unrecognized Consequences of Telecommunications Technologies

Harvey Brooks
Benjamin Peirce Professor of Technology and Public Policy Emeritus
John F. Kennedy School of Government
Harvard University

INTRODUCTION

In this chapter I shall be concerned with the application of information technologies (IT) broadly rather than only with telecommunications in the narrow classic sense of public telecom networks connecting discrete subscribers. It is by now conventional wisdom that the boundaries between telecommunications technologies and other applications of IT are so fuzzy that the distinction has lost most of its meaning. The classic notion of telecom blends imperceptibly into other embodiments of IT, including computing, office technology, and modern manufacturing technology; including the integration of design, production planning, materials handling; inventory management and market feedback into what has come to be known as computer integrated manufacturing (CIM) (Jaikumar & Bohn, 1986, pp. 169–311; Ayres, 1987). The telecom network is increasingly dominated by switching processes and is increasingly software driven. Information processing services can be made integral to the network itself or located at the terminals. CIM involves the marriage of customized software and relatively standard hardware and the controlled distribution of instructions to selected parts of the production process—almost indistinguishable from telecom, especially when different parts of the process may be widely separated. The cost of information transmission has become increasingly independent of the distance over which it is to be transmitted, making the *com* part of telecom much more significant than the *tele*.

Thus, in talking about the social impact of telecom technology, it is no longer meaningful to restrict the discussion to those functions that used to be performed by the public-switched network. Hence, we shall have occasion in this chapter to talk about manufacturing and broadcasting, and not just the transmission of voice or data between subscribers to public networks, because it is the intimate

17

intertwining of these functions that is the largest source of social impact, and which results in the most important implicit ordering of the social values and interests served by the technology.

TECHNOLOGY AND SOCIETY

A second general point that needs to be made by way of introduction is that there is an almost unconscious and implicit tendency to talk about the impact of technology upon society, as though the influence were unidirectional. In fact, the point of view of this chapter is that technological determinism is not a good window through which to view the potential social impacts of an evolving technology, and this is especially true of the proliferating family of technologies which we label IT. Thus, we insist, the flow of causation is not only from technology to society but equally or more from society to technology. One reason for this is that the social impacts of technology are almost always mediated through social support arrangements that grow up in association with the implementation and diffusion of any new technology. In this sense technological innovation is in practice almost always *sociotechnical*—a complex intertwining of social and political innovation with technical change. One consequence of this is that there is a much wider choice among many possible social arrangements or supporting systems than is generally recognized when prospective social impact is under discussion. The physical characteristics of the technology, and even the properties of the ancillary technologies that have to be developed to make it viable on a large scale, provide constraints, but not a strait jacket, on the institutional options society can choose in deploying and using the technology. At the same time the evolution of the physical embodiments of the technology is shaped by the social and cultural context in which they are embedded, even if the original technical idea (like the laser or superconductors or even nuclear fission) first appeared in an entirely technical context.

To be sure, the choices made by society in the past on how to utilize a new technology have often been made in a narrow and time-bound context shaped by existing, but transitory, social and political norms without consideration of their longer term implications, either for society or for the future directions of evolution of the technology itself. It is in the choices among immediately feasible social supporting systems that, more often than not, current (and often transitory) social values and preferences become embedded and have much more durability than was envisioned when the choices were made (Brooks, 1980, pp. 65–91). This social determination often goes unrecognized, because the values which shape its evolution are implicit. Thus, archaic social values and preferences may embedded in technological infrastructures long after the social and moral climate in which the technology originated has been completely altered.

It can be argued, in fact, that the more "advanced" the technology, the more flexibility it is likely to allow in the way society might be organized to take advantage of it. For example, the "hard" automation of the Ford assembly line heavily determined the organization of the work force and the principles of "scientific management" which became the dominant paradigm for the organization of manufacturing in the early 20th century, pioneered and carried furthest in the U.S. Yet the design of the assembly line was itself influenced by the nature of the then available work force, consisting as it did of relatively unskilled immigrants coming from a social and political background in which they were accustomed to taking orders from people with better education (Brooks & Maccoby, 1989). It is to be noted that the scientific management paradigm never achieved the complete dominance in Europe that it did in the U.S., partly because of the lack of a continent-scale market, but also because of different original social and political choices. Manufacturing technology and organization in America thus embodied elements of a mechanistic view of human behavior and motivation long after these notions had been replaced in other areas such as politics and education.

But with modern computerized manufacturing systems a much wider variety of work organizations has become possible, ranging from highly authoritarian to much more democratic and participative. The precise degree to which this is so is a matter of some debate at present. Much of the evidence is mixed. Again it is to be noted, however, that, with essentially the same production technology, a rather new industrial paradigm involving greater work force participation, compatible with today's technology, has emerged in Japan as compared with the U.S. (Kenney & Florida, 1988, pp. 121–158).

Another example of the flexibility associated with information technology is the wide national variation in the way radio and television broadcasting and programming are funded and regulated in different countries, and the consequent differences in the nature and content of the programs, expressing different political, economic, and cultural values. Yet in all these diverse systems the physical technology is virtually identical and has been adopted as standard everywhere almost as soon as it became available.

When TV was in its embryonic stages prior to WWII, there was much speculation, but little agreement, on how programming would be supported, although its potentially high cost compared to radio was often believed to be an obstacle to advertising support (Sarnoff, 1968). Yet in the U.S. a remarkable synergism developed between the technologies of mass production and the marketing of consumer goods on an unprecedented scale through the electronic media. In effect, programming was supported by a small tax on consumer goods, but because of economies of scale in mass production, the "tax" was probably more than offset by the price reductions made possible by ever larger markets. Thus the tax may have paid for itself in terms of the reduced prices that mass merchandizing on TV in combination with mass production technology made possible (Florida & Feldman, 1988, pp. 26–34). Thus, the social supporting

systems of broadcast TV reinforced and diffused the values of a nascent con-
sumerist society in America much more than in other countries. At the same
time, of course, this system of financing programming insured emphasis on the
values and tastes of popular culture. To the extent that programming succeeded
in also disseminating "high culture" as well—which it did even through the
commercial networks—this was probably cross-subsidized out of the revenues of
the popular programs, with an eye cocked toward the federal licensing system for
access to broadcast frequencies.

Yet the supporting system for the broadcast media in the U.S. was mixed in
the values that it supported. The electromagnetic spectrum was treated as a
scarce resource whose use was licensed to qualified private parties by the federal
government, with the right to use a portion of the spectrum limited by the
degree to which it served a rather vaguely defined "public interest." This "public
good" dimension moderated the values of the mass marketplace in programming, but probably not to the extent hoped for by the designers of the system.
The "equal time" doctrine also was an attempt to moderate the monopoly in the
control of information by particular private interests created by the scarcity of
spectrum. It is no accident that growing abundance of spectrum created by new
technology is leading to questioning, and some relaxation of, the equal time
doctrine.

Other ways of dealing with the scarcity of spectrum have been proposed and
debated but never actually tried. One such proposal would have auctioned slots
in the spectrum to the highest bidder and subsequently permitted such slots to be
bought and sold in a market (Meckling, 1967–1968). One may speculate
whether, if demand for spectrum had grown up in a time in which a swing
towards free market ideology such as we have seen in the last decade had been
taking place, a quite different system for the regulation of the spectrum might
have emerged, as compared with the 1930s, when the free market was in relative
disrepute because of the Great Depression.

At the other extreme from the lightly regulated U.S. system of local private
monopolies was the BBC, created as a regulated national monopoly supported by
a tax on the audience—the owners of receivers. As a result the character of
programming was much less influenced by the audience than in the U.S., where
it was virtually dictated by the comparative measurement of audience size—the
"ratings." In fact the rating system itself is an interesting example of an "ancillary technology"—in this case a social technology—which helped determine
the nature of programming (Brooks, 1982). This was not an exact measurement,
as indicated by the consternation that developed when one audience rating
company introduced a new "people meter" audience rating technology that
permitted the more accurate recording of the actual time spent watching particular programs by randomly selected viewers (Deutsch, 1987).

In Britain, even after commercial competition was permitted, the revenue
stream to the BBC was not directly affected by whether the viewer or listener

actually used his or her receiver to receive the programs of the BBC or of commercial channels. Thus, programmers for the BBC had unusual freedom to consult their own tastes and values—"elitist" relative to popular standards—in the design of programs.

In the rest of the world one finds a diversity of mechanisms of support for programming with a variety of mixes between "public good" and "private good" value emphases as well as differing views of the public interest. Some believe, however, that the advent of the small and inexpensive satellite dish will eventually abolish the influence of governments over programming, because of the ability of transmissions to transcend national boundaries (*The Economist*, 1988).

With the introduction of public television in the U.S. in the early 1970s, some features of the BBC were imported into the American system, but U.S. public television maintained an eye for the ratings to a much greater degree than the BBC. Moreover, as the tax subsidy for the system has declined, more and more market-like elements have crept into the programming.

A somewhat similar variety has emerged in the public telecommunications systems of different countries, with different organizations, industrial structures, rate-making and regulatory systems, and different implicit social goals. For example, in most countries the telephone network has been combined with the post office into a public agency, with the relatively profitable telecom system being used to cross-susidize the highly labor-intensive and increasingly inefficient postal system. However, the rapid internationalization of public telecommunications in the last two decades has placed increasing stress on the nationalistic organization of telecommunications infrastructures, with ultimate consequences that are hard to foresee (Borrus et al., 1985).

THE SHIFT TO MARKET-DRIVEN SYSTEMS (BROOKS, 1988)

The most important development of the last decade or so has been the rapid shift in the United States, and a somewhat less rapid but unmistakable shift world-wide, from a planned telecommunications infrastructure governed primarily by the goal of universal service, to a system increasingly driven by market forces dominated by the needs and demands of the largest users of telecom services. There is considerable room for argument about the relative degrees in which this shift has been caused by a sea-change in the political climate towards public regulation as compared to evolution in the nature of the technology itself. Mayo (1985, p. 23), for example, has pointed out that today's technology "is so rich that it can do more things than society might find useful" so that "marketing resources are required to sort out innovations, to contain the scope of development, and to focus investment on applications that will win in the marketplace." In other words, in this view, the shear richness of the new technology menu, and the increased intertwining of previously distinct technologies (such as data processing and transmission), have probably rendered the traditional tightly managed telecommunications infrastructures of all industrial countries obsolete. Some form of "pluralization" was inevitable even in the centrally planned economies (Starr, 1987), let alone the Western market economies.

On the other hand, telecom was neither the first nor the only regulated industry to be deregulated in the U.S., and there seems to be a good deal more than a technological imperative underlying the current developments. They are occurring in all countries, though admittedly partly in imitation of the U.S.

Whereas the pace and direction of technological innovation in telecoms were once determined mainly by systems engineering applied to the goal of expanding channel capacity and decreasing unit cost per bit of information transmitted, they are now tending to be driven by the requirements and demands of a few of the largest telecommunications users—generally large multinational corporations, the federal government, and other large organizations that are centrally managed but dispersed geographically, generally world-wide (Borrus et al., 1984). As a result the emphasis in innovation has been shifting away from the transmission and switching infrastructure towards increased sophistication of equipment and "value-added services" on the customer's premises.

The question, not yet answerable, is whether this new market-driven system will automatically ensure a telecommunications infrastructure that is optimized from the standpoint of the social values now held by a majority of America's opinion leaders, with its emphasis on "fairness" and egalitarianism. Will the system, for example, continue to meet the needs of less affluent households, small businesses, and levels of the public sector below the federal and the international?

In many ways the predivestiture Bell System was an ideal case in microcosm of a successful "planned economy." It produced a single, homogeneous, standardized product—plain old telephone service ("POTS") without terminal frills. The design of the network came as near as possible to being an exact science, the maximization of a well-defined objective function at the lowest cost consistent with a specified degree of reliability and response time and provision for future expansion. This was possible as long as the requirements of large users could essentially be met merely by increasing the number of plain old telephones within the organization, so that a large, complex organization could simply be treated as an aggregation of individuals with the same needs. Thus the old Bell system was in a position to satisfy all users with the same basic technology at the subscriber's premises. All the sophistication, which depended largely on the total number of users, regardless of their relation to each other, could be incorporated in the public network, where it was largely invisible to the public, in other words, in the production system rather than the product (Bode, 1971).

A major justification for the "old" telephone system was "technical integration": maintenance of the technical integrity of the system as new technology and new user services were incorporated into it. In the eyes of critics this justification was largely a rationalization to protect a monopoly that was increasingly anachronistic, but in the postdivestiture system the question remains whether the necessary technical integrity can still be maintained in the face of much greater diversity both in the sources of technology and in the requirements of users. More specifically, can the benefits of system-wide standards, quality control, and compatibility of different vintages of very long-lived and long-depreciation-life equipment be preserved in the new system while realizing the advantages of more experimentation and a higher rate of introduction of new user services, entailing much higher rates of obsolescence, and hence higher depreciation costs, in parts of the system?

To what extent will the new system entail compromises with the original ideal, first envisioned by Bell himself and later codified by Theodore Vail, of a universal service available at the lowest possible price compatible with a flow of resources sufficient to insure the continued growth and integrity of a continent-wide, and eventually a world-wide, network? Can the higher rate of innovation in a market-driven system offset from the standpoint of all customers the decreased attention in the system as a whole to users with less market power than the principal customers? In particular, will the residual regulation of the regional companies still assure protection of small users by insuring the continued growth and upgrading of the public-switched network without unacceptable cost increases to these customers? The original ideal of a regulated monopoly was predicated on the belief that there were large "positive externalities" associated with the widest possible dispersion of customer connections—social gains that could not be fully captured in the prices charged to the individual customer. This externality was used to justify a variety of cross-subsidies between different types of services, for example, a cross-subsidy from high density traffic routes to low density routes and dispersed users, and from sophisticated high value-added services to minimal basic services. In the United States even the idea of a "lifeline service," in effect a free or extremely low-cost entitlement to a minimum level of basic service, subsidized by the revenues from the rest of the network, came to be seriously discussed, though never widely adopted. What began as a private good thus increasingly came to be viewed as a public good or entitlement for all households regardless of ability to pay.

However, as it became technically possible to provide more and more sophisticated services on the users' premises beyond the boundaries of the public switched network, the needs of different kinds of users became increasingly differentiated, and pressures from both users and alternate equipment suppliers (other than Western Electric) became increasingly strong—in many cases irresistible. As the shift to market-driven competitive provision of differentiated services occurred, the depreciation periods for many kinds of equipment shortened, and the incorporation of accelerated depreciation rates into the tariff structure might tend to raise prices for basic service.

A new, but closely related, issue arose when independent dedicated corporate and government communications systems began to bypass the basic public network in the high density traffic corridors which had previously helped to subsidize the system as a whole, again necessitating higher prices to the smaller users who had no choice but to remain connected to the network.

There is little question that the opening up of the network to competition will tend to increase the rate of innovation, the variety of services offered, and their customization for the needs of increasingly differentiated users, but there is a question whether this innovation will be tend to be shaped largely by the needs of only the largest users, and how much the "trickle down" from these sophisticated innovations will benefit small users as well, especially small-business users. One answer may be the formation of intermediary organizations capable of aggregating the needs of many small users of a particular type of service and thus in effect becoming major surrogate users. The situation would then be much like that of the large mail order houses for ordinary manufactured goods that could dictate manufacturing specifications for the benefit of small customers if there were sufficiently many of them whose needs were sufficiently similar. Could this provide the best of both worlds?

Also, what objective function is in effect being maximized by the new mixed, partially market-driven and partially regulated system, and how will the overall configuration and performance parameters of parallel regulated and unregulated systems taken together differ from those of the old regulated system? Part of the problem here may be the difference between the short term and the long term and the question of whether long-term trends in the evolution of the system can be discerned early and surely enough to adjust the overall economic and political environment to redirect the system in a timely manner.

DISTRIBUTION OF INFORMATION POWER

A subject of intense international debate has been the question of whether "information power" will tend to become more concentrated and less universally accessible as a result of the proliferation of new technologies. On the one hand, the incorporation of more and more complex information processing capabilities in the terminal equipment of those who can afford it may tend to concentrate information power in large organizations and in major urban centers. On the other hand, the increasing abundance of channel capacity may tend in the opposite direction by making information services constantly cheaper. The role of intermediary organizations may be especially important here. This issue is of particular concern to the Third World, but is also a concern among cities and regions within both industrial and developing countries. There is a deep question of the degree to which public policy can be used to influence the accessibility of information power without damping the rate of innovation to an unacceptable degree.

What will be the influence of the future configuration of communications systems on the comparative economic advantage of various types of urban centers? Will telecommunications, probably in conjunction with air transport, reinforce the present hierarchy among urban centers ranging from world-scale economic and communications–transportation nerve centers such as New York and Tokyo to local niche cities (Noyelle & Stanback, 1983; Hanson, 1983)? Or will the abundance of channel capacity make possible many more centers with roughly equal connectedness to the world economy? There is a tendency for relative communication costs to become virtually independent of distance. How important will this be in determining the pecking order among localities, or can this effect be offset by economies of scope and scale in customer-premises equipment? How could this be influenced by the degree to which sophisticated and versatile information processing capabilities are deliberately embedded in the public network as opposed to the location of equally sophisticated but more specialized information processing capabilities installed on customers' premises?

The Japanese, in their visions of a future information society, look forward confidently to the equalization of information access both among regions and among different segments of the population, lessening the dominance of their large urban hubs such as Tokyo and Osaka (*Urban Innovation Abroad*, 1985). They speak of the information society as a force for democratization and greater public participation and are systematically linking their plans for regional economic development to the public deployment of telecommunications technology, building experimental cities that were likely to be built anyway in the normal course of economic growth in such a way as to incorporate the latest telecommunications technologies with the deliberate objective of testing their usefulness in a fully developed rather than an evolutionary situation. An example was Minato Mirai 2, a new city of 190,000 residents near Yokahama. According to Japanese proponents, this new city "will be a resort showplace and will feature wide-band communications system, videotext, and interactive systems." This is all part of the Japanese "vision" of becoming a model "information society" in the next century. The tone is one of unabashed technological optimism, and the strategy is that of creating integrated experiments to test the viability of holistic telecommunications concepts in wholly new cities (NIRA, 1985).

One may well ask how much of this Japanese optimism is mere technocratic hype and how much is truly indicative of the Japanese infrastructural future. As a country seriously lagging in its infrastructural development during its long-sustained period of export-led growth, Japan may be poised for an enormous, publicly led investment in infrastructure dominated by the exploitation of advanced information technology in all its forms and in very imaginative ways.

The question is whether this largely public investment will be made in ways that will reinforce and enhance the democratic, participatory vision embodied in their publicly announced plans. With characteristic technological optimism the Japanese are aggressively creating sophisticated public networks in the expectation that these will lead to democratization of access to information services and a reduction in the social differences arising from differential access. They appear to believe that better access to a common information base will enable lower level employees and ordinary citizens to take more responsibility for decisions with organization-wide or community-wide implications based on their more detailed and practical knowledge of operations at their own level, while at the same time having better access to and understanding of the strategic vision animating overall organizational or community decisions. Whether this vision can be realized in practice is an especially pertinent question in the light of the strong strain of large conglomeration and centralized technocratic planning in the modern Japanese political economy. Can this tradition nevertheless become the instrument for promoting more democratic and participatory values through the selective exploitation of technological opportunities (Kenney & Florida, 1988)?

With the exception of France, whose approach has some kinship to that of Japan, the view of Europe tends to be more that the new technologies give advantages to large organizations and a few communication and transportation hubs with world-scale access to information networks. This, they feel, requires some offsetting public policies on an ad hoc basis, but they profess no grand public vision.

There must be a considerable measure of choice between these visions, but the more democratic and egalitarian one may require a much higher level of public investment and collective market intervention than seems likely in view of present world-wide trends toward greater reliance on markets. There is also the question of the extent to which public intervention can be carried out more indirectly in ways that simulate the functioning of markets by "internalizing" social costs and benefits in the pricing of different kind of services.

The U.S. seems inclined, for the time being at least, to let nature (or the free market) take its course, allowing much of the evolution of the technology to be dictated by competitive vendors, and systematically rooting out cross-subsidies wherever they are found. The problems are viewed as so complicated that the public cannot be much involved. The public, in turn, is perceived to be interested mainly in basic telephone services and diverse entertainment options at the lowest possible cost. Some knowledgeable observers of the American scene see telecommunications and IT more broadly as reinforcing all the inequalities that have developed recently in American society, stratified largely along the dimension of educational level. According to this view, both the mastery of and access to IT increase with educational level and tend to amplify initial inequalities resulting from differences in formal education (Downs, 1985, pp. 26–34).

CENTRALIZATION VS. DECENTRALIZATION

There is also an ongoing debate about the implications of information technologies for the nature and structure of organizations. On the one hand, telecoms have greatly increased, at least in principle, the ability of a central headquarters to monitor and control far-flung, indeed world-wide, activities, thus making possible highly centralized bureaucracies of even greater scope and power than those that exist at present. On the other hand, the possibility is also present for creating more horizontal network-like organizations consisting of quasiautonomous nodes capable of coordinating their activities voluntarily by virtue of instant access to a common data base and to each other by means of telecoms.

In other words, it is uncertain whether information technology will be used to enhance the capacity of a few "big brains" to coordinate and control the activities of more and more people with a high degree of functional specialization to a common end, or whether it will increase the capacity of numerous "little brains" to concert and coordinate their activities in a way that is equivalent to, but more efficient than one big brain that cannot possibly keep all the information necessary for central decision making at the forefront of its attention simultaneously and continuously, no matter how powerful the processing capabilities of its information terminals. The potential for centralized control may be inherently limited by the problem of information overload of the central decision maker, whose decisions are ultimately dependent on semi-intuitive judgments rather than the ever more refined management of ever larger volumes of information. Some commentators put great faith in the power of artificial intelligence techniques to overcome this problem of information overload, but others suggest that, while information processing can change the relative roles of "formal" and "informal" processes in the manipulation of information, there is still a limit to the decision power of a single brain. The effectiveness of centralized management is inherently limited by the integrating capacity of the individual brain.

Those who favor "flat," or network-type, organizations believe that better access to a common information base will enable lower level employees to better identify with organizational goals and to take more responsibility for decisions with organization-wide implications based on both their more detailed and practical first-hand knowledge of operations at their level and on better access to and understanding of the strategic vision of the organization as a whole which, in a more hierarchical organization, it would be assumed they did not need. In hierarchical organizations such decisions would be pushed to the top on the assumption that only the boss would have the "big picture" and be in a position to insure consistency across a wide range of functional areas.

On the other hand, coordination among many people in a horizontal organization presumes a large measure of prior agreement on the proper decision rules informed by common values and goals. To the extent that these decision rules are formalized, individual responsibility, and hence motivation, may be diluted. In fact formalized, explicit decision rules must be sanctioned either by hierarchical authority or by strong peer pressures, so that, in either case, real discretion and individual autonomy is absent. To the extent that many individuals must synthesize their intuitive judgments, the problem of consistency among these judgments becomes more critical, even though the amount of information that has to be processed in the decision process at any one node is less than at the center. There is also a question of the extent to which the flat organization model can be implemented in a society as individualistic in its traditions and comprised of as diverse cultures as that of the U.S.

Nevertheless, in this situation, I think there is real choice, an opportunity to deliberately exploit technology to design organizations which are either more bureaucratic and hierarchical, or more democratic and participatory, according to the basic values of the society as a whole or to the "corporate culture" of the specific large organization in which it is embedded. The choice between centralization and decentralization is thus not predetermined by the technology. Some decisions will no doubt remain centralized, but decentralized organizations may be more efficient, because responsibility can be better matched to detailed operational knowledge at different levels. The learning process for organizations, however, may be longer and more difficult in pluralist, individualistic societies.

FLEXIBLE SPECIALIZATION?

Piore and Sabel, Hirschhorn, and others have argued that modern communications-intensive manufacturing and office technologies make possible a new vision of dispersed "flexible specialization" in which manufacturing systems can be profitable with much shorter production runs and can produce a larger variety of products with the same capital equipment (Piore & Sabel, 1984; Hirschhorn, 1984). This, in their view, greatly reduces the potential benefits of scale economies, and makes possible more rapid and less costly responsiveness to changes in customer requirements and in the competitive environment. In many service as well as manufacturing industries there has been a marked trend away from fine division of labor, with corresponding increases in discretion and responsibility at lower levels of the organization (Baran, 1984).

In looking at what is actually happening in the world, we find an amazing degree of diversity. The vision of world-wide coordinated operations has certainly been realized in some multinational corporations, but, even here, with considerable devolution of responsibility to lower levels. At the same time it has often been possible for small companies in narrow market niches to secure access to world-wide markets, something they could not have done without modern information technology and telecommunications. For example, the highly fragmented apparel manufacturing and distribution system in the United States has been able, through information technologies, to greatly reduce the time between customer ordering and delivery of a finished product, resulting in dramatic reductions of stock and the costs of carrying large inventories as well as of having to offer slow-selling products, produced before the market could be properly assessed, at low margins or at a loss. There is some evidence that this exploitation of information technology may be restoring some of the competitiveness of apparel manufacturing in the U.S., because of the greater responsiveness to customers made possible by sophisticated IT (OTA, 1988).

However, these configurations are not exclusively determined by economic efficiency narrowly defined. This is because of the growing importance of workforce commitment to the long-term effectiveness and adaptability of the organization. This commitment is a function of the social values of the individuals comprising it. An organization that looks more efficient on paper may prove to be less effective in practice, because it is poorly matched to the social and ethical values of its members (Brooks & Maccoby, 1989).

THE DEVELOPING WORLD

An interesting and important issue related to the centralization–decentralization dichotomy is how information technology may affect the division of labor between the industrialized and the developing world. Until recently it has appeared that telecommunications, combined with air transportation, has facilitated the physical separation of design and manufacturing, the subcontracting of the more labor-intensive steps in manufacturing to low-wage countries, and the servicing of customers over great distances. This is particularly true for standardized products that meet the needs of diverse customers with a single design. The division of labor has thus tended to conform to the product cycle theory of the diffusion of technological innovation (Vernon, 1966, pp. 190–207). But the newer forms of flexible manufacturing, and the increasing emphasis on inventory savings and "just-in-time" manufacturing systems in "fragile/lean" organizations, as well as tighter integration of design, manufacturing, marketing, and customer services, may be shifting the advantage back to the developed countries (*Newsweek*, 1984, p. 178; *Fortune*, 1984, p. 152; Castells & Tyson, 1988, pp. 55–95).[1] Also, as automation proceeds, the fractional labor content of many

---

[1] Cf. especially pp. 58–66 for a discussion of the "comparative advantage reversal" hypothesis and its evaluation for the particular case of microelectronics (Castells & Tyson, 1988).

goods and services has become so small that labor costs are often no longer a significant consideration compared to inventory costs, round-the-clock utilization of capital, the sophistication of the local transportation and communications infrastructure, and the availability of the highest quality skilled workers and technicians to service and maintain production equipment. Thus, dispersed production that includes LDCs may entail higher inventory costs and costly shut-downs of production, or delayed deliveries to customers because of equipment breakdowns, transportation delays, or customs red tape—additional costs no longer offset by cheap labor.

Also, as many consumer goods become more differentiated and customized, the importance of locating manufacturing plants and service centers close to customers increases. The balance of locational considerations may, of course, come out differently for different products and for different industrial sectors. Hence, broad generalizations are difficult. It is also less and less certain that the equation will always be balanced in purely economic terms as the industrialized countries, particularly the U.S., become more sensitive politically to loss of employment perceived to be due to off-shore sourcing of components and subsystems. Moreover, the labor-cost advantages of many developing countries are likely to be a wasting asset as gaps in living standards gradually close and shortages of skilled labor appear in the countries that are the most successful competitors (*The Economist*, 1988, pp. 16–17). Whereas access to cheap labor may be a transient competitive advantage, being the earliest to start down the learning curve with the latest manufacturing technology may generate an advantage that grows with time, while the labor-cost advantage erodes over time, as the Japanese have found. The final answer here is not clear, as it depends on a delicate balance between so many factors.

THE POLITICAL PROCESS AND GOVERNANCE

One of the principal impacts of communications upon both politics and the governance of organizations has been the speeding up of many of the important transactions and processes involved, with correspondingly reduced time for deliberation over many important decisions. The implications of this for the operation of democracies, as well as for the governance of many types of organizations, may be subtle and difficult to foresee. I would argue, for example, that the effective functioning of a democracy requires deliberation. When decisions are made too rapidly, whether by referenda, politicians' responses to opinion polls, legislative action, or executives transmitting orders to distant locations on the basis of information fed back instantly from the field of operations, the quality of decisions is likely to deteriorate. Decision makers, or the public, are more likely to react emotionally or irrationally to incomplete or inaccurate information, or to superficial "images" of reality generated by the media, rather than the actual reality. Abundance of information often serves to becloud rather than clarify the true situation. The more information there is, the greater the opportunity for each person to selectively perceive only that information that fits his or her preconceptions or policy preferences. By making instant reaction more feasible in many circumstances, modern communications may contribute to a deterioration of the quality of decision making in a democracy, despite the fact that they may have given more people better access to more accurate information. The greater fidelity of the information channel may not be sufficient to offset the poorer quality of hasty mental processes and the difficulty of picking out the most relevant information from a mass of data.

We see this phenomenon in the reaction of politicians to media stories and instant opinion polls, in decisions made by high level officials in crisis situations (as in the recent shooting down of civilian airliners by both the Soviet and American military), in the reaction of the public to an overload of information about political candidates in primary campaigns, in the susceptibility of a considerable segment of the public to TV evangelists, and in hundreds of other ways.

A few years ago there was a major public debate about the desirability of banning the computerized projection of election results by the electronic media during elections because they could influence patterns of voter turnout, particularly in national elections. France actually has such a ban.

The tyranny of the 15-minute time slot in national television news may be another example of how time constraints may result in inadvertent distortion of information, or open up the possibility for deliberate distortion or manipulation of news or public opinion.

The stock market "meltdown" of October 1987 launched a national debate over whether the combination of modern communications and fast computers has introduced dangerous and unrecognized instabilities into financial markets. More generally, instant communication among financial managers world-wide may reinforce a tendency toward herd-like decisions, as well as opening financial markets to virtually undetectable forms of insider manipulation for personal gain. Even in the absence of actual abuses or illegal activities, public trust in the fairness of markets may be undetermined merely because of the theoretical possibility of undetectable insider manipulation. More generally, the instantaneous and simultaneous availability of the same information (or misinformation) to thousands of people may reduce the independence of opinions and judgments about current events.

Modern telecommunications will soon make possible virtually instantaneous public referenda on important national issues. Some technical people are enthusiastic about such possibilities, seeing them as a contribution of technology to more democratic decision making, turning a whole state or nation into a town meeting. Yet there is serious question that this is a welcome development or that this form of "real time direct democracy" would constitute a desirable form of political decision making. Citizen deliberations under time pressures are much more likely to be irrational and impulsive. Moreover, some politicians would be only too happy to avoid responsibility for their hard choices by reference to such instantaneous feedback from the electorate. Trust in the electoral system, and in democracy more broadly, would be likely to erode because of the suspicion of manipulation.

Unpredictable "information overload" among powerful executives or other decision makers may not only lead to the deterioration of the quality of decisions affecting the lives of many people, but also may give rise to various group pathologies of thinking and consequent systemic social instabilities difficult or impossible to predict.

CONCLUSIONS

It is generally characteristic of advanced technologies that there is a much wider range of choice in the social arrangements that can be compatible with the technology than was true for less advanced technologies. However, there is a risk that these choices will be made by default according to transitory values and narrow interests unless systematic thought is given to the possibilities at an early stage in the development of the technology. This may be particularly true of telecommunications and information technologies more broadly. Visions of alternative futures need to be widely discussed.

Probably the most important development in relation to telecom technologies is the shift—most radical in the U.S., but world-wide in scope—from planned monopolies of telecom towards market-driven systems. At the present time it is unclear how far this shift will go, or what its influence will be on the evolution of the technology and of the social arrangements surrounding it. Granted that it will be driven by the needs and demands of a few of the largest users, it is still not clear how compatible the resulting configurations of technology will be with the needs of smaller users. The rate of technological innovation is undoubtedly accelerating and will continue to do so, but it is uncertain how well the "trickle down" of new technology to the broader market from the most sophisticated requirements will actually serve a larger public.

A related question regarding the impact of the new technology has to do with the distribution of "information power" within society—e.g., the comparative economic advantage of different types of cities ranging from international communications and transportation hubs to regional centers serving specialized market niches, of large and small businesses, of big countries and small countries, of the Third World vs. the industrialized countries, of the well educated and the less well educated. The Japanese, among others, have put forward a vision of telecom as a force for democratization, for the empowerment of an increasing fraction of the population, and for the equalization of the comparative economic advantage of localities, but many doubt whether such a vision can come to pass if the evolution of systems and networks is left largely to the marketplace and there is an absence of political intervention to reflect the "public good" aspects of the information infrastructure.

While there is no question that, in the recent past, modern telecoms and transportation have facilitated the decentralization of labor-intensive manufacturing to many parts of the Third World, there are indications that the newest developments in manufacturing systems and telecoms may be nudging production back towards the industrialized countries. However, this varies so much from sector to sector that it is difficult to be certain as yet of any general trend.

Finally, the rapidity with which information can now be disseminated worldwide is raising questions about the problem of "information overload" of decision makers, whether they be the executives of large organizations or the ordinary citizen expected to form opinions on complex issues which are increasingly influential with the politicians who represent him or her in governance. It is doubtful whether the sophisticated processing by electronic means of increasing volumes of information can sufficiently reduce the minimum of judgment and intuition required to make critical social decisions. Hence, there is question of whether "information power" may be making modern societies ungovernable.

REFERENCES

Ayres, R. U. (1987, November). *Manufacturing and human labor as information processes.* IIASA RR-87-19.

Baran, B. (1984). *Insurance industry and trade strategies* (Draft Report submitted to OTA, project on "Technology and the American Economic Transition"). Berkeley, CA: Berkeley Roundtable on the International Economy (BRIE), University of California.

Bode, H. W. (1971). *Synergy: Technical integration and technological innovation in the Bell System.* Murray Hill, NJ: Bell Laboratories.

Borrus, F. B., Bar, F., Cogez, P., Thoresen, A. B., Warde, I., & Yoshikawa, A. (1985). *Telecommunications development in comparative perspective: The new telecommunications in Europe, Japan and the U.S.* Berkeley, CA: Berkeley Roundtable on the International Economy, University of California.

Borrus, M., Bar, F., Warde, I., with Millstein, J., & Cogez, P. (1984). *The impacts of divestiture and deregulation: Infrastructural changes, manufacturing transition, and competition in the U.S. telecommunications industries.* Berkeley, CA: (Project for U.S. Congress, Office of Technology Assessment) Berkeley Roundtable on the International Economy, University of California.

Brooks, H. (1980, Winter). Technology evolution and purpose. *Daedalus,* pp. 65–91.

Brooks, H. (1982). Social and technological innovation. In S. B. Lundstedt & E. W. Colglazier (Eds.), *Managing innovation: The social dimensions of creativity, invention and technology.* Elmsford, NY: Pergamon Press.

Brooks, H. (1988). Reflections on the telecommunications infrastructure. In J. Ausubel & R. Herman (Eds.), *Cities and their vital systems: Infrastructure, past, present and future.* Washington, DC: National Academy Press.

Brooks, H., & Maccoby, M. (1989). Corporations and the work force. In J. R. Meyer & J. Gustafson (Eds.), *The U.S. business corporation.* Cambridge, MA: Ballinger Publishing.

Castells, M., & Tyson, L. D. (1988). High technology choices ahead: Restructuring interdependence. In J. Sewell & S. Tucker (Eds.), *Growth, exports and jobs in a changing world economy.* New Brunswick, NJ: Transaction Books.

Deutsch, C. H. (1987, July 26). The battle to wire the consumer. *The New York Times,* Section 3 (*Business*).

Downs, A. (1985). Living with advanced telecommunications. *Society, 23*(3), 26–34.

The Economist. (1988, August 27). All the world's a dish. pp. 7–8.

The Economist. (1988, August 20). Seoul's first event. pp. 16–17.

Electronic Business. (1984, February). p. 178.

Fortune. (1984, July 11). p. 152.

Florida, R. L., & Feldman, M. M. A. (1988). Housing in U. S. Fordism: The class accord and postwar spatial organization. *International Journal of Urban and Regional Research, 12*(2), 187–210.

Hanson, R. (1983). *Rethinking urban policy: Urban development in an advanced economy* (Chap. 3). Washington, DC: National Academy Press.

Hirschhorn, L. (1984). *Beyond mechanism: Work and technology in a postindustrial age.* Cambridge, MA: MIT Press.

Jaikumar, R., & Bohn, R. E. (1986). The development of intelligent systems for industrial use: A conceptual framework. *Research on Technological Innovation, Management and Policy, 3,* 169–311.

Kenney, M., & Florida, R. (1988). Beyond mass production: Production and the labor process in Japan. *Politics & Society, 16*(1), 121–158.

Meckling, W. H. (1967–1968). Management of the frequency spectrum. *Washington University Law Quarterly, 4 & 5,* 26–34 (Special Issue on *The Radio Spectrum, Its Use and Regulation*).

Mayo, J. S. (1985). The evolution of information technologies. In B. R. Guile (Ed.), *Information technology and social transformation* (pp. 7–34). Washington, DC: National Academy Press.

National Institute for Research Advancement. (1985). *Comprehensive study of microelectronics 1985.* Tokyo, Japan: NIRA.

Noyelle, T., & Stanbach, T. (1983). *Economic transformation of American cities.* Totowa, NJ: Allanheld & Rowman.

Newsweek. (1984, March 12). p. 36.

OTA. (1988). *Technology and the american economic transition.* Washington, DC: Office of Technology Assessment.

Piore, M., & Sabel, C. (1984). *The second industrial divide: Possibilities for prosperity.* New York: Basic Books.

Sarnoff, D. (1968). Looking ahead. *The Papers of David Sarnoff* (pp. 50–51). New York: McGraw-Hill.

Starr, S. F. (1987, October). *New communications technologies and civic culture in the USSR.* Paper presented at Center for International Affairs, Harvard University.

Urban Innovation Abroad. (1985, April 4–5). *Impacts of New Telecommunications Technologies on Local Governments in Europe, Japan and the U.S.* Explored in International Symposium.

Vernon, R. (1966). International investment and international trade in the product cycle. *The Quarterly Journal of Economics, 80,* 190–207.

chapter 3
Telecommunications, Self-Determination, and World Peace

Chadwick F. Alger
Mershon Center and Department of Political Science
The Ohio State University

INTRODUCTION

The relationship between communications and peace is complex and fluid. New technology is dramatically changing the volume and velocity of information flows, and the ways in which information is collected, stored, transmitted, presented, and accessed.* New actors in world politics, from new states to grassroots movements, are giving new meanings to *peace.* As World War II drew to a close, a strong feeling emerged among those planning the postwar world that the *free flow of information* would contribute to peace by lowering the likelihood of war. But closely following on World War II, the winds of *self-determination* swept strong Third World voices into international conferences and organizations dealing with global communications who found "free flow of information" to be a somewhat misleading description of a global communications system dominated by transnational media corporations in the industrialized countries. They demand "balanced flow" among countries and regions. At the same time a global dialogue based on the meaning of peace was emerging because of dissatisfaction with limiting the definition of peace to the absence of war. From the perspective of those taking a broader view of peace, a *balanced flow of communications* can be viewed as an attribute of peace.

To those accustomed to defining peace as the absence of war and to assuming that *free flow of information* is an obvious prerequisite for peace, Third World perspectives on peace and communications may appear as distorted and politicized uses of familiar words whose meaning is obvious. But to those employing a broader definition of peace, and those who believe *balanced flow of communica-*

---

* I am grateful to Roger Coate and John E. Fobes for comments on the first draft.

tion to be an attribute of peace, the impact of the technological revolution on global communications has made it necessary to rethink old assumptions about peace and global communications.

Some witnessing growing global conflict over communication may be driven to despair as they note that a prime means for pursuing peace—increased communication across national boundaries—has itself intensified conflict, particularly between the North and the South. On the other hand, we are inclined to believe that the present conflict over global communications can be viewed as an indispensible laboratory for defining the kind of global communications structure that will be responsive to the needs of a diversity of regions, interests, and cultures. Understanding of the significance of this dialectic to peace requires comprehension of both dramatic changes in communications technology in this century and of changes in the international context in which these changes have occurred. The former is extensively covered by other chapters in this volume. We shall briefly review the latter.

FREE FLOW OF INFORMATION AND PEACE

Representatives of 44 countries gathered in London in November 1945 to create an organization that would emphasize the role of international understanding in the attainment of peace. The French representative, Leon Blum, wished the new organization to "establish the spirit of peace in the world." The Peruvian delegate said that education must recognize the unity of the human race. The representative of Yugoslavia asked for reeducation of the people of the Axis countries and of "ourselves." The Indian delegate asserted that education must refashion itself to reshape the world. "In short, the delegates at London called for better education to fashion better men for a new life in a democratic society" (Laves & Thomson, 1957, p. 6). In pursuit of these goals they approved a UNESCO Constitution which began with words familiar to many: "That since wars begin in the minds of men, it is in the minds of men that the defenses of peace must be constructed." But few are cognizant of the two following paragraphs which give explicit meaning to this opening phrase:

> That ignorance of each other's ways and lives has been a common cause, throughout the history of mankind, of that suspicion and mistrust between peoples of the world through which their differences have all too often broken into war; . . .

> For these reasons, the States Parties to this Constitution, believing in full and equal opportunities for education for all, in the unrestricted pursuit of objective truth, and in the free exchange of ideas and knowledge, are agreed and determined to develop and to increase the means of communication between their peoples and to employ these means for the purposes of mutual understanding and a truer and more perfect knowledge of each other's lives.

In pursuit of free flow of information, UNESCO promoted a series of international agreements. The Universal Copyright convention adopted in 1952 standardizes international copyright procedure for literary, scientific, and artistic works, including books and other writings, music, records, films, painting, and sculpture. The Agreement on Importation of Educational, Scientific, and Cultural Materials, approved in 1950, eliminates duties on books, newspapers, magazines, educational films, and recordings and works of art. In 1954 an agreement came into force for Facilitating the International Circulation of Visual and Auditory Material, Materials of an Educational, Scientific and cultural character. In response to a UNESCO initiative, participants in GATT reduced import duties on educational, scientific, and cultural materials. To promote the free flow of information, UNESCO has worked for the reduction of postal, freight, and telegraph rates, and has recommended measures to facilitate transmission of press messages (1955) and abolition of censorship (1953) (Laves & Thomson, 1957, pp. 133–136). UNESCO also developed a world system for exchange of scientific documentation (UNISIST), which later expanded to become the very successful General Program of Information. Although many constraints on "the free exchange of ideas and knowledge" still remained two decades after the founding of UNESCO, for many, free flow of information had become an integral part of their vision of a peaceful world. But self-determination and new telecommunications technologies would soon produce rapid changes that would call into question this vision of peace.

TRANSFORMATION OF THE INTER-STATE SYSTEM

In September 1960, 16 new African states entered the United Nations, signifying that colonial empires were beginning to yield to aspirations for self-determination. In December of 1960, the General Assembly passed the "Declaration on the Granting of Independence to Colonial Countries and Peoples" with 89 voting yes, no negative votes and nine abstentions. The attainment of rapid independence by most remaining colonies produced a future UN of 159 members, including 51 from Africa, 35 from Asia, 33 from Latin America and the Caribbean, 7 from Oceania and 33 from Europe and North America. Thus in only a few decades the UN was transformed from an organization dominated by Europe and North America to an organization with over 100 Third World members.

Spurred by the growing gap between the rich North and the poor South, both bilateral and multilateral aid programs were created that provided technical assistance, loans, and grants intended to spur national development in the South. Nevertheless, the gap continued to grow. In response Third World analysts claimed that the cause of their difficulty was the dependence of their economies on those of the industrialized states. They charged that Third World countries were doomed to remain economically inferior to the industrialized states until the inequitable international economic structure was changed. A new organization in the UN System, the UN Conference on Trade and Development (UNCTAD) became the arena for crystallizing Third World demands for a New International Economic Order. In the historic Sixth Special Session of the UN General Assembly in May 1974, the Third World achieved acceptance of a Declaration on the Establishment of a New International Economic Order (NIEO), and an accompanying Programme of Action. In order to achieve equity in international economic relationships, the Third World was asking for stability in commodity prices, indexing prices of commodities to the prices of manufactured products, access to needed technology, national control over natural resources, and international regulation of transnational corporations. Lack of responsiveness by the North led to a demand for Global Negotiations for implementing a NIEO. But protracted debate in the General Assembly over a number of years brought little response from the North, with the United States, the United Kingdom, and the German Federal Republic leading the resistance.

Frustration over lack of responsiveness to demands for Global Negotiations on a NIEO contributed to a growing Third World view that people in the industrialized countries did not understand their predicament because of the structure of global communications systems, as exemplified by the fact that the press agencies with global reach all had their headquarters in Europe or North America. Indeed, the dependency of Third World countries on foreign press agencies mirrored their economic dependency.

The natural arena for efforts to challenge the existing global communications order would be UNESCO. In 1980, "Principles for a New World Information and Communication Order" (NWICO) were approved by the UNESCO General Conference. A major emphasis of the principles is "plurality of sources and channels of information," respect for the right of all peoples to participate in the exchange of information, and "elimination of the imbalances and inequalities which characterize the present system." As means for achieving these goals the principles call for "elimination of the negative effects of certain monopolies, public or private, and excessive concentrations," and a reminder that "the freedom of journalists in the communications media [is] a freedom inseparable from responsibility." It was in pursuit of greater "responsibility" that some governments represented in UNESCO had throughout the 1970s called for the licensing of journalists, although this very controversial proposal was never approved, nor even voted on. Nevertheless, the licensing of journalists issue received wide reporting and editorial comment in the United States, although newspapers reporting this issue rarely informed their readers of the reasons for the emergence of the NWICO.

WHAT IS PEACE?

A remarkable global dialogue on the meaning of peace has been in process over the past 25 years, centered in the United Nations and related nongovernmental organizations, in the transnational peace research community, and, increasingly, in grass-roots movements. The point of departure for the dialogue was the definition of peace as the absence of war. Third World scholars in particular challenged this definition by pointing out that far more lives are lost because of inadequate food, clothing, shelter, and medical care than are killed by wars. Thus peacelessness has two causes: One is the *direct violence* associated with war, and the other is *structural violence* (or indirect violence) that is caused by the way that societies, and relations between them, are organized. It follows that there are two kinds of peace. The peace achieved by stopping direct violence is called *negative peace*. The peace achieved by the creation of more just societies, and more just relations between them, is called *positive peace*.

Two different kinds of relationships between negative peace and positive peace are important. First there are causal relationships, as when social injustice leads to direct violence, or when war or threats to security create conditions that contribute to structural violence. Second, there are political relationships produced by the fact that people whose greatest fear is direct violence tend to prefer policies that give priority to elimination of direct violence; and people whose greatest fear is structural violence tend to prefer policies that given priority to elimination of structural violence. An implication of these differences in priority is that in many cases a political compromise must be made between the two points of view.

A particularly dynamic aspect of the global dialogue on the meaning of peace has been ever-deepening understanding of the meaning of positive peace. Often cited is the Pope's assertion that "development is another word for peace." But certain development strategies imposed upon unwilling people have been declared to be structural violence. The same is said of development that destroys local culture or the local environment. Dramatic examples are given of the peacelessness of polluted streams, wells, and fishing grounds, and of development policies that destroy the traditional customs of everyday life. For the sake of simplicity one could say that the progressive definition of positive peace is involving the definition of human rights, economic well-being, and ecological balance. Significant are the connections between the three, as in the case of economic well-being (as defined by others than the recipient of economic "benefits") that is imposed from the outside, thus violating human rights.

Following this line of reasoning one can understand how communications that contribute to the destruction of local culture, communications that tend to flow only in one direction without the possibility of reciprocal influence, and communications that gain entrance without the possibility of local opportunity to reject them can be considered the exercise of structural violence and thus contributory to peacelessness. How ironic that technology that makes it possible for people at one spot on the globe to share information about themselves with any other spot on the globe may not contribute to the peace that free flow of information was supposed to enhance. Much to the contrary, it may, and often has, produced peacelessness.

DEMAND FOR A NEW WORLD INFORMATION AND COMMUNICATIONS ORDER (NWICO)

We have delineated two points of origin for the NWICO. One is the lack of responsiveness of the industrialized countries to the demand for a NIEO, a strategy for overcoming the growing gap between the rich and the poor of the world by creating a more equitable international economic structure. The other is the potential that communications structures have for creating peacelessness. There is yet another point of origin, the struggle of Third World peoples for self-determination. There is no doubt that the collapse of the overseas colonial empires have transformed the globe, but in many cases "political" independence has not been accompanied by economic independence. Indeed, the NIEO can be viewed as a strategy for extending self-determination into the economic sphere. At the same time, the NWICO can be viewed as a demand for communications independence. Self-determination will not be completed until all three are achieved.

But the self-determination issue is even more complicated, because most of the Third World states are themselves creations of colonial administrators that are preventing the fulfillment of desires for self-determination, or at least greater autonomy, by hundreds of nations. Many of these nations not only suffer the peacelessness produced by communications from the West, but also are victims of communications policies of Third World states. As Dov Ronen so effectively argues, "decolonization" not only leaves unresolved the "ethnic self-determination" needs of many peoples in Africa, Asia, and Latin America, but also those in Europe and North America (Ronen, 1979, pp. 39–52). Communication within groups, as well as between groups, is a critical factor in the quest for cultural identity.

One can trace the roots of the NWICO to as early as April 1959, when the Committee of Peaceful Uses of Outer Space raised concern about the impact of international broadcasting on national sovereignty. In 1960 the UN Economic and Social Council requested UNESCO to survey world press, radio, film, and TV development. Attention began to be focused on the control of information by transnational news agencies, on charges that these agencies distort news in response to their Western clients, and on criticism of the lack of news flow between Third World countries.

In 1961 UNESCO began encouraging the formation of regional organizations by national news agencies in Asia, Africa, and Latin America, and the effort gained momentum when the fourth conference of Heads of State of Non-Aligned countries meeting in Algiers (1973) recommended the creation of a nonaligned news pool. This was preceeded by a UNESCO resolution addressing the problem created by the intrusion of satellite communication into sovereign territories without permission. At the same time the heads of state noted the linkage between national cultural identity and indigenous control of communication in Third World countries. As these initiatives of the Third World gained momentum, they often obtained the support of the Soviet Union and its allies, although the Soviet definition of the problem often differed from that of the Third World. Thus, a political struggle developed, with Western governments, supported by transnational media corporations and the Western press on one side, and Third World and Eastern bloc governments on the other.

In the Nairobi meeting of the UNESCO General Conference in 1976 the Western countries responded with an initiative of their own. The U.S. delegate acknowledged that the Third World had some legitimate claims and offered "American assistance, both public and private, to suitably identify centers of professional education and training in broadcasting and journalism in the developing world" as well as "a major effort to apply the benefits of advanced communications technology—specifically satellites—to economic and social needs in the rural areas of developing nations" (Nordenstreng & Schiller, 1979, p. 197, cited by Pendakur, 1983, p. 398). In the next meeting of the UNESCO General Conference in 1978, a compromise text on "Declaration on the Media" was drafted by delegates from the U.S., Germany, Italy, Tunisia, and, occasionally, Poland. Following the adoption of this resolution, a U.S. Department of State spokesman commented that the Declaration "not only is stripped of all language implying state authority over the mass media, but also includes positive language of freedom of information" (UNA, n.d., cited by Pendakur, 1983, p. 398). Significantly, this declaration, acceptable to the U.S., followed a 1972 Byelorussian SSR proposal for a Declaration on the "uses of the media" and the rejection of a first draft of a "Declaration on the Media" in 1974.

These events led one scholar to observe:

> The US seems to have deflected the intense pressure from the Soviet and Non-Aligned nations successfully by resetting the agenda of UNESCO debates, preserving its own ideological position of "free flow" of information in the text and placing the whole battle in the context of technology transfer, thereby shifting the focus of power away from the coalition of Socialist and Non-Aligned nations to Washington. Thus the existing core-periphery structure of international communication was preserved at least for the time being. (Pendakur, 1983, pp. 398–399)

The Nairobi meeting also directed the appointment of an international commission to study communications problems. This 16-member group, with members from capitalist, socialist, and nonaligned states, was headed by Sean Mac-Bride of Ireland and completed a voluminous report, *Many Voices: One World* (International Commission for the Study of Communications Problems, 1980) that "should be seen as a negotiated document between the NWICO proponents and opponents" (Pendakur, 1983, p. 399). This became the basis for a New World Information and Communication Order resolution adopted by the General Conference of UNESCO in Belgrade in 1980 which is notable for its effort to combine notions of (a) free flow, (b) balance, (c) both freedom and responsibility for journalists, and (d) both respect for cultural identity and right to communicate with the rest of the world. The resolution reads:

The General Conference considers that

(a)   This new world information and communication order could be based among other considerations on:
  (i)   Elimination of the imbalances and inequalities which characterize the present situation;
  (ii)   Elimination of the negative effects of certain monopolies, public or private, and excessive concentrations;
  (iii)   Removal of the internal and external obstacles to a free flow and wider and better balanced dissemination of information and ideas;
  (iv)   Plurality of sources and channels of information;
  (v)   Freedom of the press and of information;
  (vi)   The freedom of journalists and all professionals in the communication media, a freedom inseparable for responsibility;
  (vii)   The capacity of developing countries to achieve improvement of their own situations, notably by providing their own equipment, by training their personnel, by improving their infrastructures and making their information and communication media suitable to their needs and aspirations;
  (viii)   The sincere will of developed countries to help them attain these objectives;
  (ix)   Respect for each people's cultural identity and for the right of each nation to inform the world about its interests, its aspirations and its social and cultural values;
  (x)   Respect for the right of all peoples to participate in international exchanges of information on the basis of equality, justice and mutual benefit;
  (xi)   Respect for right of the public, of ethnic and social groups and of individuals to have access to information sources and to participate actively in the communication process;
(b)   This new world information and communication order should be based on the fundamental principles of international law, as laid down in Charter of the United Nations;
(c)   Diverse solutions to information and communication problems are required because social, political, cultural and economic problems differ from one country to another and, within a given country, from one group to another. (Pendakur, 1983, p. 408)

THE CAMPAIGN AGAINST NWICO IN THE UNITED STATES

Despite this movement toward a definition of NWICO that responded to both the Western desire for free flow and the Third World aspiration for balance, a strong offensive that had developed against the NWICO and UNESCO was sustained. Given the self-determination struggles of the United States in the 18th and 19th century, it is in some respects surprising that the United States government led governmental resistance to the NWICO. At the same time there are grounds for surprise that the U.S. media led media resistance. Indeed, the media in the United States were still attempting to extricate themselves from British and French domination of worldwide communications as recently as the 1940s. In 1942, Kent Cooper, executive manager of the Associated Press, in his book *Barriers Down,* charged:

> in precluding the Associated Press from disseminating news abroad, Reuters and Havas [French] served three purposes: (1) they kept out Associated Press competition; (2) they were free to present American news disparaging to the United States if they presented it at all; (3) they could present news of their own countries most favorably and without it being contradicted. Their own countries were always glorified. This was done by reporting great advances at home in English and French civilizations, the benefits of which would, of course, be bestowed on the world. (Cooper, 1942, pp. 76–77)

Cooper also made this more graphic complaint: "So Reuters decided what news was to be sent from America. It told the world about Indians on the warpath in the West. . . . The charge for decades was that nothing credible to America was ever sent" (Alfian, 1976, pp. 10–11). These criticisms mirror those of advocates of a NWICO today.

Furthermore, as with Third World governments today, the United States government directly entered the fray. In January 1946 a State Department broadcast reported the following views of Assistant Secretary of State William Benton:

> The State Department, he said, plans to do everything within its power along political or diplomatic lines to help break down the artificial barriers to the expansion of private American news agencies, magazines, motion pictures, and other media of communications throughout the world. . . . Freedom of the press—and freedom of exchange of information—is an integral part of our foreign policy. (US Department of State, 1946, quoted by Schiller, 1975, p. 77)

Thus a longer historical perspective on NWICO would help the United States to understand that there is more than one approach to free flow of information, depending on where you are located in the world power structure. And it would help the United States to recognize that Third World policies are spurred by the same problems faced by the United States not long ago.

According to Pendakur, "the major counter offensive against UNESCO" came from the World Press Freedom Commission (WPFC) consisting of 31 media organizations, including Time, UPI, AP, and the National Association of Broadcasters. Headed by the president of the *Omaha World Herald*, the WPFC "appears to represent mainly the media owners' interests and claims to be financed entirely by their donations" (Pendakur, 1983, p. 405). Opposing both the MacBride Report and the Belgrade resolution, they organized a widely publicized conference at Talloires, France in 1981, entitled "Voices of Freedom." The conference produced a declaration pledging "cooperation in all genuine efforts to expand the free flow of information and called on UNESCO to "abandon attempts to regulate news content and formulate rules for the Press" (*Time*, June 1, 1981, p. 82, cited by Pendakur, 1983, p. 405).

The U.S. Congress enthusiastically joined the movement. In June 1981 Senator Dan Quayle of Indiana introduced a measure opposing UNESCO and the NWICO. Before it was passed by a voice vote, Senator Daniel Moynihan introduced an amendment reading that "it is the consensus of Congress that the United States should withhold from its contribution to UNESCO our share of the money UNESCO chooses to spend implementing its misguided New World Information Order." The amendment passed with 99 favorable votes. In September of 1981 Representative Beard proposed in the House of Representatives a resolution that required the Secretary of State to report to the House by February 1 of each year whether UNESCO had any directives curtailing press freedom. President Reagan wrote a letter to the Speaker of the House in support of Beard's proposal, and it passed 372 to 19 (Pendakur, 1983, p. 406).

On December 28, 1983, the United States announced its intent to withdraw from UNESCO, "without formal consultation with its Western Group partners" (Coate, 1988, p. 1). Following the obligatory waiting period, the Reagan administration terminated U.S. membership at the end of 1984. In withdrawing the United States gave three primary reasons: politicization of UNESCO's programs and personnel, promotion by UNESCO of statist theories, and budgetary expansion and poor management. But there are puzzling elements in the U.S. government policy process leading to U.S. withdrawal. Commenting on a U.S./UNESCO Policy Review begun in 1983, with contributions from all federal agencies with interest in UNESCO's programs, Coate concludes: "if one is searching for support for withdrawal in these interagency assessments of UNESCO, it is hard to find" (Coate, 1988, p. 26). At the same time, the U.S. National Commission for UNESCO, composed of a broad array of national educational, scientific, cultural, and communications organizations, held a conference devoted to a "Critical Assessment of Relations with UNESCO" in June 1982. "The five working groups at this conference concluded unanimously that the U.S. should stay in UNESCO and increase its participation and exert a strong leadership role" (Coate, 1988, p. 7).

Coate asserts that a February 1983 report to Congress by the Department of State concluded that "UNESCO's programs for the most part contributed to

U.S. foreign policy goals and particular U.S. educational, scientific and cultural interests," and that "vigorous continued U.S. participation was essential to protect U.S. interests" (Coate, 1988, p. 17). At the same time, "By most accounts, conditions in UNESCO in those problem areas cited in justification of withdrawal had improved markedly in recent years" (Coate, 1988, p. 5). Another scholar reached an even stronger conclusion: "The United States, it seemed, had walked away just when it was winning the fight" (Stevenson, 1988, p. 165).

Why then did the U.S. withdraw? Coate (1988) concludes that official policy of the U.S.

appeared to be one of getting the US and other member states to quit the organization, not a policy of reforming UNESCO. . . . US policy toward UNESCO had been captured by a small group of ideological zealots, who seemed to know or care little about UNESCO and its problems. (p. 5)

It is reasonable to conclude that the media campaign against the NWICO provided the special circumstances making U.S. withdrawal from UNESCO feasible without substantial public controversy. Pendakur calls the content of this campaign "a caricature of the NWICO," that is, by presenting the NWICO as primarily concerned with imbalances in news flow. But, says Pendakur,

it is about information in a broad sense of the term, which includes business and scientific data, military and strategic information and advanced delivery systems of computers and satellites, all of which impinge on the sovereignty of nations. It is not merely a fight over press freedom and democracy as the WPFC would have it, but it is a broad debate covering the transnational power structure and the Third World's attempt to alter it. (1983, p. 405)

In other words, the issue is self-determination!

For the most part the American public was never provided information about the broad context of the Third World NWICO campaign, about its relationship to the NIEO, about resistance to U.S. withdrawal by government departments and nongovernmental organizations involved in UNESCO programs, and about the important of UNESCO educational, scientific, and cultural programs to U.S. interests. It is the deepest of ironies that, in this case, the U.S. media itself prevented the free flow of information about the communications conflict in UNESCO—the kind of information that would have enabled the American public to understand the difficult problems created by transformations in telecommunications technology. The irony goes even deeper when we note that an eventual impact of the U.S. media's failure to fully inform the public on the NWICO was to make it possible for the U.S. government to refuse to take part in NWICO debates in UNESCO—the appropriate arena for multilateral debate and exchange of ideas on a vitally important global issue. Here again, the U.S.

media has contributed to inhibiting the free flow of information between states that are parties to this conflict.

ARE PEACEFUL COMMUNICATIONS POSSIBLE?

As early as 1979, Ithiel Pool attempted to strike a middle path that in spirit resembles the UNESCO Belgrade declaration in 1980:

> Believers in free discourse cannot accept a policy for the mass media that subjects them to a censorship designed to prevent them from violating the norms of culture, but neither should humane men disregard the pain that rapid change imposes on men whose life cycle is no different than it was millennia ago. (1979, pp. 146–147)

He proposed that the United States work with Third World countries "to narrow the gap in information resources between the industrialized countries and the LDCs [Less Developed countries]." Therefore, he believed it to be "very much in the U.S. government's interest to subsidize a flow of communications by satellite from the developing countries to the rest of the world and also of communication by the United Nations. . . . The United States could propose, for example, that the UN establish a worldwide TV network that would distribute programs originated all over the world" (Pool, 1979, p. 149). Thus, rather than basing policy on a self-centered, and self-serving, definition of free flow, Pool asked for a policy that would indeed facilitate greater reciprocal flow of information worldwide. Importantly, Pool pointed out that there are those in society who have a predisposition to support this broader view of free flow of information:

> There is fortunately a coincidence between the true interests of development and the ideological and moral predispositions of at least one important group in society. Scholars, scientists, practitioners of the intellectual arts tend to be among the strongest advocates of free flow. To liberal intellectuals the right to communicate is sacred; the doctrine of the free flow of information is a principle not be to compromised. (1979, p. 153)

Writing only one year later, in *Geopolitics of Information: How Western Culture Dominates the World*, Anthony Smith sees the revolution in new micro-electronics and new telecommunications systems as a "threat to independence. . . . that could be greater than was colonialism itself." The news media have the power to penetrate more deeply into a "receiving" culture than any previous manifestation of Western technology.

> The results could be immense havoc, an intensification of the social contradictions within developing societies today. In the West we have come to think of the 2500

communication satellites which presently circle the earth as distributors of information. For many societies they may become pipettes through which the data which confers sovereignty upon a society is extracted for processing in some remote place. (Smith, 1980, p. 176)

He asks that policies responsive to this "new industrial revolution . . . be conducted more with a view to its global implications than to the short-term interests of the transnational companies and quasi-monopolies which dominate these fields" (p. 176).

Yet Smith also has scathing words for Third World advocates of the NWICO:

Perhaps the greatest weakness in the list of demands which makes up the New International Information Order has been its lack of conception of the primal value of press freedom (and of intellectual freedom as a whole). In essense, the order was formulated by the wrong people in the wrong way, although much of the sentiment supporting them has been genuine and even in certain respects liberating; but seldom can the charter of a great political cause have been so mean in spirit, so ungenerous in sentiment, so obsessively petty, so insistent upon the obligations of others and so niggardly in ascribing difficult duties to its own adherents. (p. 176)

Concerned that "the inequalities between world sectors could become irretrievably gross," Smith sees the need for "real international action" (p. 176). On the other hand, he makes the international action he would like to see look exceedingly difficult, if not impossible, when he notes that "the problems are so deeply rooted in history and in geology of human attitudes that one cannot expect to see a reversal of these inequalities without major shifts in world power greater than those which have yet taken place" (p. 174).

In *Electronic Colonialism: The Future of International Broadcasting and Communication*, Thomas L. McPhail argues that, in some respects, presumed differences between the Western press and the Third World press to be found in conflict over the NWICO are overdrawn. He notes that "development journalism" is a notion that encourages a press committed to the development policies of Third World governments. On the other hand, he observes that the Western mass media assume the ideological role of protecting, perpetuating, and enlarging the role and influence of the capitalist system. "Indeed, editorials and feature columnists continually call for less government control and less regulation in order that market forces be allowed to control the economy." Thus, he says that the Western press is a "development press" too, and has in fact successfully developed itself into an ideological arm of the capitalist and private enterprise system. He concludes that development journalism, or development communication, is "in essence . . . the underlying rationale for all media and communications systems," whether they be the Western free press model, the Socialist model or development journalism as applied to the Third World. With respect to the West, he notes that

> both free flow and free press are designed in such a way as to continue and enhance the economic benefits that are derived by the owners and organizations who staunchly defend such claims and traditions. . . . The commercial enterprise itself, over time, is therefore developed into an ever larger conglomerate whether it be Gannett, ATT, Capital Cities, or IBM. They have been involved with development—their own—for decades. (McPhail, 1987, p. 296)

In terms of the socialist model, they seek to develop and reflect the central party line. With respect to the Third World, they want news media and communication technologies that will serve "their objectives rather than what they perceive to be the more narrow economic objectives of the Western technology or software suppliers" (McPhail, 1987, p. 296).

Based on this array of perspectives, as well as the evidence brought to bear in this volume, there appears to be little doubt that the introduction of computers and satellites into the systems, of digital movement of all forms of data and information, will have as fundamental an impact on the organization of humankind as the era of "discovery" that led to the development of overseas colonial empires. At the same time, it seems clear that humanity has not yet begun to find ways to come to grips with this fundamental transformation. Furthermore, intelligent response is made particularly difficult because the media itself has the ability to prevent a wide and searching examination of global communications issues, whether in the East, the Third World, or the West.

As Anthony Smith has said, without "real international action," the future looks bleak, both with respect to self-determination and peace in the broadest sense of that term. For this reason, the NWICO debate in UNESCO had been a very hopeful development, and the movement in UNESCO toward some kind of accommodation of "free flow" and "balanced flow" advocates was encouraging. Yet the capacity of a small cabal in the U.S. government—supported by a self-serving campaign of the American media and aided by the apathy of a globally uneducated American public—to withdraw the U.S. government from this significant debate is ominous.

On the other hand, a weakness of using the UNESCO arena in attempting to reach accommodation on global communications issues is that it is composed of representatives of states who tend to represent the interests of states as they see them. This has tended to produce a grossly oversimplified debate. One problem is frequent failure to illuminate the self-determination struggles of many national, ethnic, and cultural groups which are directed against the information policies of their own states, as well as global communications. Another problem is that the "North–South" and "East–West" debates in UNESCO often obscure the common interests of all the people of the world. Perhaps this could be overcome by much more active participation of nongovernmental groups—both national and transnational—having both concern about the "free and balanced flow of information" and concern about public participation in the formulation of national and global communications policies.

At the same time, there is a need for a host of experiments in transnational communication that begin to illuminate the potential for communication that is really fulfilling of the needs of all involved. The Pan Pacific Education and Communication Experiment by Satellite (PEACESAT), linking terminals in sixteen different countries in the Pacific, would seem to be a relevant example. PEACESAT is a network of fully independent terminals, each able to determine for itself its participation in project exchanges. Begun in 1978, PEACESAT has been used for educational experiments involving teachers, administrators, planners, and parents; health and medical experiments have included reviews of medical and surgical problems and control of epidemics; social and community activities include youth groups and community service clubs jointly planning cooperative international projects. As described by Anthony Hanley (1978, p. 717) PEACESAT seems to be contributing to the developing of communications networks supportive of self-determination and peace in the Pacific.

CONCLUSION

In 1945 the fathers of UNESCO were correct in their understanding that free flow of information can contribute to peace. Writing over 30 years later, in the light of a dramatically changed world, Colin Cherry reached a similar conclusion in *World Communications: Threat or Promise*:

> The values of global communication . . . lie in the contribution which the various systems can make towards this end, towards the practical removal of frustrations. The values lie not in their unlikely power for persuading everybody else to be like us, but rather in putting our various distinct characteristics, arising from our different histories, geographies and peoples to positive value. That is, to work *through* these differences, not *upon* them, and to use these differences, which may seem to be dissent or even heresy, as a creative source of change of our own institutions. For whatever progress is, it must include a preparation for such change, as other peoples' institutions change. (Cherry, 1978, p. 204)

But in the great peace laboratory of UNESCO we have now learned that, without self-determination of all parties to communications, the results can deprive self-determination and create peacelessness.

REFERENCES

Alfian. (1976). *Some observations on television in Indonesia*. Paper presented at the Conference on Fair Communication Policy, East-West Center, University of Hawaii, Honolulu, HI.

Coate, R. (1988). *Lessons from UNESCO: Unilateralism and beyond in American foreign policy*. Boulder, CO: Lynne Rienner. (Page numbers in citations are taken from a pre-publication manuscript.)

Cooper, K. (1942). *Barriers down*. New York: Farrer and Rinehart.

Cherry, C. (1978). *World communication: Threat or promise? A socio-technical approach* (revised ed.). Chichester, England: John Wiley and Sons.

Hanley, A. (1978). Intercultural international communication. In F. L. Casmir (Ed.), *Intercultural and international communication* (pp. 717–740). Washington, DC: University Press of America.

International Commission for the Study of Communication Problems. (1980). *Many voices, one world* (MacBride Commission Report). Paris: UNESCO.

Laves, W. H. C., & Thomson, C. A. (1957). *UNESCO: Purposes, progress, prospects*. Bloomington, IN: University of Indiana Press.

McPhail, T. L. (1987). *Electronic colonialism: The future of International Broadcasting Communication* (revised 2nd ed.). Beverly Hills: Sage.

Nordenstreng, K., & Schiller, H. I. (Eds.). (1979). *National sovereignty and international communication*. Norwood, NJ: Ablex Publishing Co.

Pendakur, M. (1983). The new international information order after the MacBride Commission Report: An international powerplay between the core and the periphery countries. *Media, Culture and Society, 5,* 395–409.

Ronen, D. (1979). *The quest for self-determination*. New Haven, CT: Yale University Press.

Schiller, H. I. (1975). Genesis of the Free Flow of Information Principles: The Imposition of Communications Domination. *Instant Research on Peace and Violence, 5,* 75–86.

Smith, A. (1980). *The geopolitics of information: How Western culture dominates the world*. New York: Oxford University Press.

Pool, I. de Sola. (1979). Direct broadcast satellites and the integrity of national cultures. In K. Nordenstreng & H. I. Schiller (Eds.), *National sovereignty and international communication* (pp. 120–153). Norwood, NJ: Ablex Publishing Corp.

Stevenson, R. L. (1988). *Communication, development and the Third World: The global politics of information*. New York: Longman.

Time. (1981, June 1). p. 82.

UN Association of USA. (n.d.) *The new international information order—Who wants it? What is it? Issues of the 80s*. New York: UNA and USA.

U.S. Department of State. (1946). *Bulletin, 14,* 3.

chapter 4

Language, Global Telecommunications, and Values Issues

Johanna S. De Stefano
Professor, Graduate Program in Language, Literature and Reading,
    and Senior Faculty Associate, The Mershon Center
The Ohio State University

INTRODUCTION

With the world caught up in a communications revolution which has ushered in the Information Age, the capabilities made possible by telecommunication technologies have outstripped their thoughtful use or much in the way of policy planning regarding use. This turn of events could well be what led Dizard to state, "Success in our generation will depend on the degree to which we shape the new information technologies in accordance with human values" (1982, p. 11). There is strong feeling in many lesser developed countries (hereafter referred to as LDCs) that *their* human values are not being taken into account in the face of a constant barrage of programming and data brought to them via the technology (cf. McPhail, 1986; Schiller, 1976; Tsuda, 1986).

Further, the use of telecommunications technology impacts the use of indigenous languages in many countries, as the United States has dominance in both the hardware and programming, which is in English. Strevens (1987, p. 57) notes, "As the telecommunications revolution got under way, English became dominant in the international media, radio and TV, magazines and newspapers. . . . so, too . . . [in] . . . space science and computing technology." But English is dominant far beyond its use in the telecommunications industry. As Kachru puts it, "For the first time a natural language has attained the status of an international (universal) language essentially for cross-cultural communications (Kachru, 1983, p. 85). Or to quote Wardhaugh, "English has become the lingua franca of the modern world" (1987, p. 137), "by far the most widespread of the world's languages" (1987, p. 128).

Because of this dominance, "issues of language differences seldom enter the consciousness of mass media producers and planners" (Tonkin, 1984, p. 73). I have to agree that, even 5 years later, the "issues of language differences" are not forcefully raised or widely acknowledged as issues of importance. This chapter will raise these issues to consciousness and scrutiny, exploring possible effects the spread and use of English may have on values held in countries at the "receiving end" of telecommunications technology. The so-called McBride Report for UN-ESCO on the New World Information Order stated the issue this way:

> the use of a few so-called world languages is essential in international communica-tions, yet it poses sensitive questions concerning the individuality and even the political and cultural development of the countries. (International Commission for the Study of Communication Problems, 1980, p. 51)

This chapter will explore the role that English plays in the interaction between values such as sovereignty and development in LDCs and telecommunications technology as offered by developed nations, and as such is the only exploration of this intersection that I'm aware of in the literature.

LANGUAGE AS WINDOW, MIRROR, OR KALEIDOSCOPE?

An examination of the possible effects of using English while trying to develop and yet maintain sovereignty in an LDC must begin with questions about the relationship between language and culture. Is language a mirror? Does English, for example, reflect the British and American cultures and their values, implying their acceptance? Does an Indian look in the mirror and see someone who is Indian but more stereotypically British than the British, a "brown sahib" who identifies with the "white sahib" (Kachru, 1982a, p. 40)? Do English users in LDCs "reflect" all that can attach to the language? There are those who feel that's virtually impossible to avoid.

Or is English a window? Can a language be transparent, a more neutral vehicle for communication, a tool to be used by whoever wishes to use it? A window on the larger world? There are those who feel this is possible.

Or is it a kaleidoscope, giving a "fractured," distorted image of the message? Some argue that this is the case, that the English language conveys values which are not clearly articulated, so inevitably leads to distortion in the cultures which learn English as a second or foreign language. It's something akin to the planet Bizarro in the Superman comics—the same and yet not.

ENGLISH AS A WORLD LANGUAGE

These important questions about the impact of English on the values of modernization and sovereignty in LDCs begin to be clarified by an understanding of the current status of English as a world or international language, or, as it's some-times called, an LWC—Language of Wide Communication. How did it get to the point that people on a global basis run the gamut from welcoming it as a unifier in a country fraught with internal dissention to vilifying it as a robber of local dreams and ideals, a desecrator of indigenous cultures, a spreading stain on the face of the world?

First, English is widely recognized as a major international language, in fact the primary "link language" in the world today (Fishman, Cooper, & Conrad, 1977, p. 56), and has become an "instrument of access" for people almost everywhere (Weinstein, 1983, p. 90). This link language status is illustrated by the following example. The Educational Testing Service's TOEIC (Test of English for International Communication) program director Stupak notes, "a Tokyo-based company may have plants in Indonesia, Thailand and Singapore, and offices in the United States, France and Mexico. . . . English is probably the language in which all these offices and operations communicate with each other" (*ETS Developments, 1988*).

English is this international language precisely because more people speak it as a second, third, fourth, or even fifth language than as a first language. Consequently, there are many more than the 300 million native speakers for whom this is the case, leading to a total "well in excess of 700 million" (Wardhaugh, 1987, p. 135). In fact, Wardhaugh states this to be the smallest figure one can arrive at.

> There seems to be little doubt that about one-fifth of the world's population of better than five billion people has reason to use some English almost every day. (1987, p. 135)

These figures still put English behind Mandarin Chinese in numbers, but Chinese does not have the link language status, being spoken largely on the mainland of China by people who are mostly native speakers. On the other hand, if one takes Strevens's figure of 1.5 billion total users of English, then it is ahead of Chinese (1987, p. 56).

Other languages such as Spanish, Hindi, Arabic, and Portuguese (in Brazil) may soon surpass English in terms of numbers of native speakers (Wardhaugh, 1987, p. 135), but none so far have the international language status of English. English has an official status, either by law or by use, in more countries in the world than any other language. It is the single official language in 25 countries and a co-official language in 17 more countries. Its nearest competitor is French, which is official in 19 countries and co-official in nine (Wardhaugh, 1987, p. 135). And when we think of other powerful countries such as Russia and Japan, we do not see similar patterns of language spread.

Definitely, the British empire began the spread of English, with one of its strongest branches now American English. It also spread to countries with well-established indigenous populations such as India, a variety of African countries, and many others where it became firmly entrenched as the language of government, at least of higher education, if not the entire educational system, of the legal system, and so on. Much of this use still exists today. India recognizes English as one of its official languages, actually linking the country together, as indigenous languages seem even more fraught with emotion. Although only about 3% of the population is actually bilingual in English and Indian languages (Kachru, 1983, p. 52), approximately 50% of the books published in India are in English, while much of government and higher education is conducted in English. Other examples abound, as in Malaysia, Singapore, Nigeria, and elsewhere.

ENGLISH IN SCIENCE AND TECHNOLOGY

But even though the tide of empire has receded, English has become ever stronger as a link language. We can explain that in part by the passing of the mantle from Britain to the United States and its vast multinational corporate empire, whose influence has given rise to the term *neocolonialism*. As Wardhaugh states, "The 'centre' of English has shifted across the Atlantic," where "the North American variety benefits from its associations with science, technology, the media, and raw political, economic, and military power" (1987, p. 129).

U.S. control over much of the mass media has been noted above, and that control is usually exerted by companies whose business it is to create and sell news, information, programs, films, and so on. But other U.S. businesses have had their impact as well. Currently the language of most international banking is English. So too in the airline and shipping industries for international communications. To sum it up, Weinstein states:

Ignorance of English eliminates one from the competition for the highest positions in banking, aviation, and international commerce in non-English-speaking countries. IBM employees from France to Korea are sent to language school by the corporation to learn American English. (1983, p. 68)

McPhail (personal communication, May 6, 1988) indicated that IBM's presence in the spread of English is even more pervasive, in that the manuals and technical instructions accompanying their computers are in English. So even in Quebec, which is militantly francophone, French-speaking technicians may have to use computer manuals written in English, although there is movement to write such manuals in French. Furthermore, many Quebecois play Apple computer games—in English.

Thus, in the minds of many, English is viewed as *the* language of science and technology. Weinstein comments that "English [is] now considered more appropriate for scientific communication than other languages, even those with rich scientific traditions" (1983, p. 21). Wardhaugh details this hegemony; he notes that, in 1980, 72.6% of the articles listed in the *Index Medicus* were in English. The next largest figure was 6.3%, for Russian! He goes on to assert, "It is difficult to understand how a scientist who cannot read English or who does not have immediate access to good translations from English can hope to keep up with current scientific activity" (1987, p. 136). Tsuda, who largely bemoans the spread of English, elaborates on this problem, asserting that the dominance of English in science is so complete that it prevents scientists who speak other languages, not only from keeping up with the literature, but also from participating fully in scientific meetings and journals, from actually making their work known, or from getting the attention it may deserve (1986, p. 2). In a slightly different vein but demonstrating the same sort of hegemony, Kaplan (1982, 1983) and McCrum, Cran, and MacNeil (1986) note that 80%–85% of all information stored in world-wide databases is either in English or at least abstracted in English.

This view limits what English is often seen as being useful for. Fishman, Cooper, and Conrad, in their seminal volume on the international growth of English, conclude, "English is considered to be more acceptable for technology and natural science use than for political and social science use, and . . . it is least acceptable of all for local humanistic and religious purposes" (1977, p. 124). Certainly this is what Shaw (1983) found when he conducted a study of Indian, Singaporean, and Thai students' attitudes toward English. Many of the top-ranked reasons for studying the language were based on what is called *instrumental* rather than *integrative* motivation. It's seen as a needed language in business and education, both of which tend to rest on science and technology. But these students did not feel knowledge of English would "make me a better person" (Shaw, 1983, p. 23), which is considered to be an integrative motivation. Thus English could be characterized as being thought of as a language in which to get things done, e.g., the language of development which flows from scientific discoveries and the use of technology.

THE PROBLEM OF LINGUISTIC DETERMINISM

English is widely accepted as the major scientific and technical language of the world today, the major link language for development, and is the language carried via power technology forms—telecommunications—which are considered to play an important role in development. What might the language link to both the production and use of technology mean for people in LDCs and the values they hold? In order to begin to answer this question, we must determine whether English is a mirror, a window, or a kaleidoscope. Is it a "loaded weapon," the mirror and kaleidoscope categories, as many claim, or can it be relatively neutral, like a window?

As one might expect, this is a hotly contested issue, with much written and widely disseminated by believers in linguistic determinism or so-called linguistic relativity. This position was articulated in the writings of Edward Sapir (in Mandelbaum, 1956), an early social scientist, and the most popularly by Benjamin Lee Whorf (1956), so it's sometimes called the Sapir-Whorf or Whorfian hypothesis. Simply stated, the hypothesis asserts that language strongly influences thought. Sapir said, "forms predetermine us to certain modes of observation and interpretation" (in Mandelbaum, 1956, p. 7).

As these "forms" Sapir indicated include grammatical form as well as vocabulary, the hypothesis tends to equate verb tense with time. Past tense is about actions happening in the past, e.g., past time, and future tense is about actions to come. For example, Chinese has no verb tenses, so the hypothesis would assert that Chinese people have difficulty conceiving of time. Yet Chinese speakers obviously are aware of their past and their future, with a written history of around a 4000-year span. As Weinstein says so well, "Grammar does not maintain a vicelike grip on the mind. People perceive differences and similarities even if their particular language does not provide them means to express them" (1983, p. 25).

In terms of vocabulary, if a language has many, varied terms for events, things, etc., then the Whorfian hypothesis asserts it's easier to think about them than in a language with a less-developed vocabulary in those areas. For example, in English there are multitudinous terms for short pointed pieces of metal made for sticking in things and holding them together, for example, nails, tacks, brads, staples, etc. However, English speakers, as a group, are not preoccupied with these items, although specialists *can* speak very specifically about them if they need to. The connection between language and thought is just not as simple as the hypothesis would have us believe.

However, linguistic determinism is a very popular and prevalent position, one which holds that at best language is a mirror and at worst a kaleidoscope, operating on both thought and culture. So when one learns English as a foreign language and metaphorically looks in the mirror, one will see, instead of a Nigerian or a Malay, an English person or an American with those values, beliefs, and so on. As a kaleidoscope, English is held to distort a people's thinking about their own culture, to "fracture" their world view. On the other side of the coin, "[a native] language makes a people, and a people without pride in its language is dispossessed of its national pride. Preserving one's language is preserving one's culture" (Hofman, 1977, p. 277). This was said by a Shona-speaking teacher trainee in Zimbabwe reacting to the continued use of English there.

Others make essentially the same point. For example, Laitin (1977, p. 162) hypothesized:

A country's choice of one national and official language over another meant the choice of one behavior over another; therefore, the continuing position of European languages as official in most countries is a partial explanation for the persistence of colonial values and institutions in independent Africa.

Tsuda (1986, p. 49) states:

> The English language is not merely a medium, but represents the soul, ideology and way of life of the English-speaking people. Therefore, the world-wide spread and use of English may result in the world-wide diffusion of the ideology and cultural practices of the English-speaking nations.

Later he assets that it does *not have* to be the case but actually *is* the case in that speakers of other languages, by losing their choice of languages to speak through *having* to speak English, actually lose their freedom to "choose *how* they think, live, build their nations, and so on" (p. 56). English, he maintains, transmits its culture and "creates changes in human consciousness" (p. 57).

These assertions about the effects of a language on thought and culture are widely held. Fishman (1982, p. 17) gives us the example of recent language planning in the Philippines in which English as a medium of education was restricted to math and natural sciences, the "ethnically less encumbered" subjects. The rest of the curriculum was to be taught in Pilipino, the national language, which is considered ethnically grounded, even though it too represents only one ethnic group among the many in the Philippines. Another example: approximately 4 years ago the government of Malaysia began to use Bahasa Malaysia, the relatively newly proclaimed national language, for university-level education which had been in English, a leftover from days of colonial rule. Cohen (1988), reporting in the *Chronicle of Higher Education* detailed the problems now being faced at the university level because of that decision, not the least of which is a major dearth of advanced materials in Bahasa Malaysia. But she concluded the article with:

> Despite Malaysia's problems in replacing English as the medium of undergraduate instruction, many academics here think the change is politically and pedagogically sound.
>
> Says Mr. Aziz, the University of Malaysia economist: "You cannot have a cultural identity, a personality, in someone else's language." (p. A30)

This implies that, when you study in English, for example, rather than your own language, by making that shift you wouldn't see *yourself* in a mirror, as your personality and identity are "given" you by the other language. Wardhaugh says just this, that "there is a *widespread* (emphasis added) belief that a shift in language often brings about a shift in identity" (1987, p. 5).

THE DENATIONALIZATION OF ENGLISH

The important question is whether or not this widespread belief in linguistic determinism is justified in the face of other views such as the "alternative 'value-free' view of English in the world" as Wardhaugh puts it (1987, p. 132), or the "English as a window" argument. There is strong evidence for the validity of this latter view due to the existence of a wide variety of local Englishes. They have arisen or are arising all over the world, are very different from one another, may be becoming standardized, and are definitely indigenous (DeStefano, 1985, p. 119). They began as English impacted by the surrounding local languages, giving rise to what is called Indian English, Philippine English, and West African English, to name only a very few. Indian English has strong Hindi influences in it; Philippine English has influences from Tagalog, the indigenous language which with some modifications has been raised to the status of the national language and renamed Pilipino; and West African English has influences from at least Hausa, Yoruba, and other languages indigenous to Nigeria and other countries in the region. Some of these Englishes have become locally standardized, with a literature as well. They are legitimate types of English which fit the non-British, non-American, usually non-Western environment. They are forms of English whose users have made them their own.

Even if English is not heavily impacted by another language, as is the case with the Irish variety which has not been changed by exposure to Irish Gaelic, a Celtic language, it can evidently serve others' needs. The Irish provide an excellent example of this. Often called England's oldest and most troublesome colony, they increasingly speak English at the "expense" of Gaelic. In fact, Irish culture and independence are being asserted in English, not in Irish Gaelic. Their indigenous, non-English language is actually dying out, while their culture is being achieved in English (Edwards, 1985), now the language of an independent Ireland able to make its case in the "language of the oppressors."

This denationalization of English is a process seen elsewhere as well. One of my favorite examples comes from English teaching materials from the People's Democratic Republic of Yemen. As Krishnaswamy and Aziz state, they have produced "some teaching material which is in tune with their national aspirations and values" (1983, p. 99). Note that the Yemeni is writing to an Indian— an example of English as a common, link language.

A Letter From a Pen Friend

Aden
14th October, 1978

Dear Mohan,

Thank you very much for your letter and the photographs of Hyderabad. I learnt a lot about your country and your customs from your letter. It was better than reading a book on India.

. . . you have asked me why we want to learn English. I must tell you that we were opposed to the English rulers. We said, "ya Inglizi Barra". (You English, get out.) We are not opposed to the English people or the English language. We fought against the British rulers and British imperialism. Even now we are fighting against all forms of imperialism. We are not against the people of any country. I think that a knowledge of English will help us in our war against imperialism, poverty and ignorance. . . .

(*Yemeni Reader, Book I*, People's
Democratic Republic of Yemen)

Shaw's study of the Indian, Singaporean, and Thai students' attitudes clearly illustrates the denationalization process. He reports that the three countries' students did not feel it important to study English so that they could "think and behave as native speakers do" (Shaw, 1983, p. 24).

Others writing of other continents or subcontinents reflect much the same view. Mazrui (1975), a Ugandan scholar and coiner of the term *AfroSaxon*, quotes Moorehouse as correct in asserting that English has been widely adopted in Africa as a "politically neutral language beyond the reproaches of tribalism" (in Mazrui, 1975, p. 15), and cites an estimate that around 66% of Black Africa is becoming English-speaking. Kachru, an Indian, contends that LDCs use English to "teach and maintain the *indigenous* patterns of life and culture, to provide a link in culturally and linguistically pluralistic societies, and to maintain a continuity and uniformity in educational, administrative and legal systems" (1976, p. 155; emphasis added).

If it is the case that English is being and can be denationalized and used as a relatively neutral vehicle for self-expression, then why do so many espouse linguistic determinism? I think Fishman (1977c) has analyzed the situation succinctly when he suggests that, at times of intense, conscious ethnicization in countries, there may be similar intense feelings against an LWC, a language of wider communication, which may not be only English. He has found that this often coincides with "urbanization, industrialization, modernization, and political integration efforts" (Fishman, Cooper, & Conrad, 1977, p. 119), processes occurring currently in many LDCs which are striving both to establish a national identity from among disparate groups, for example, establish sovereignty, and to modernize, the two major values I will discuss below vis-à-vis English's role in their realization.

What seems to happen, in part, is that many confuse language use with the language itself, so that both are discarded by some, while in other situations, as in Yemen, the language and its use are seen to be separate. Baxter says it well. "The *use* of English is always culture-bound, but the English language is not bound to any specific culture or political system" (1983, p. 104; emphasis added). In other words, don't mix up verb tenses with time concepts. There are different ways different languages code time concepts—via vocabulary, structure and so on. However, as language is used in a context, the combining of structure with function is not surprising.

On the other hand, the French language, whose structure is, of course, neutral, has not passed itself off in a neutral manner at all, but has been tied consciously to French culture as a matter of policy by the government. These differences between the two languages prompted Fishman to quip, "English is less loved but more used; French is more loved but less used" (1982, p. 20). That neatly sums up the major attitude toward English as a world language. It gets the job done, so to speak, without a *necessity* for cultural encumbrances even though they exist in some situations, especially postcolonial.

SOVEREIGNTY AS A VALUE

Given this sociolinguistically based analysis of the role and use of English in LDCs, what interaction might the language have with the use of telecommunications technology in helping countries achieve what they value? I've selected two major values for discussion in this chapter—those of sovereignty and modernization (via transnational cooperation). These two seem to be widely held values throughout the LDCs and merit close scrutiny, in part because of the role telecommunications technology plays in each. Schiller asserts, "for the new nations which were colonies not so long ago, the effort to create communications-cultural policies for national liberation and to satisfy the working people's needs for better material conditions of existence is no marginal item" (1976, p. 71).

Sovereignty is frequently valued against the backdrop of former colonial status for many LDCs. As colonies they often had imposed on the indigenous peoples a European culture and language. Such imposition gave rise to the "brown sahib" Kachru (1982b) described and the forerunner of Mazrui's (1975) AfroSaxon, to name only two of the many manifestations of the contact among cultures, the colonizer culture being dominant and the indigenous ones usually being subordinated.

With independence, the process of decolonization often began as a way to achieve or actualize the value of sovereignty. It's considered very important to be a sovereign state, to have a vote in the United Nations, and to support and maintain the local cultures within the new national boundaries or forge a new national identity, sometimes at the expense of some indigenous groups. Schiller goes so far to state that "a paramount concern of these states is to safeguard their national and cultural sovereignty" (1976, p. 39). Thus the world has witnessed a strong sense of nationalism arising in many LDCs. India, Pakistan, The Philippines, Malaysia, Indonesia, Nigeria, and Zimbabwe present only a few examples of this process at work.

Against these strivings, we must place the domination of telecommunications from the United States and, to some extent, from Great Britain. The question has arisen, then, about how a new nation which values sovereignty can achieve decolonization, a process felt to be necessary to assert that desired sovereignty, in the face of what is perceived to be continued domination, now via telecommunications technologies? The Prime Minister of Guyana in 1973 stated the problem this way: "A nation whose mass media are dominated from the outside is not a nation" (*Intermedia*, 1973, p. 1).

Out of these very real fears has come UNESCO's New World Information and Communication Order (NWICO), which is a policy, described elsewhere in this volume (see Alger, Chapter 3) designed to speed the process of decolonization and promote a healthy sovereignty by enabling individual countries to have more control over the media which reach them via telecommunications, as well as by other means. The NWICO does not address the role of language, but does enlarge upon the problems addressed in the New World Economic Order, as both economics and communications are clearly tied together.

However, when it's felt that sovereignty rests on the use of indigenous languages, telecommunications use presents a conundrum because of the hegemony of English in the programming, in the information and data flow, etc. But do indigenous languages make the achievement of sovereignty more possible? Tsuda argues that they do, in that the colonization process leads to "the rejection and disintegration of indigenous culture and language" (1986, p. 27). He reasons that the choice of an indigenous language helps in decolonization, in achieving sovereignty. Somalia's current leaders made such a choice. According to Laitin (1977), over 95% of the people of Somalia can understand Somali, an indigenous language. However, English emerged as the national language after independence from Britain. After a coup in 1969, the military regime which took over declared Somali the national language, proclaimed the Latin-based script (one of several competing script forms) official, and began the process of education, government, and so on in this national language. Obviously, the value of sovereignty was felt by the new rulers to be best attained through the use of Somali rather than English as the language of government, education, and other public domains. In fact, Fishman asserts, a national language is widely felt to be needed even for "mass *mobilization* along the road to modernity" (1977c, p. 331).

I have also cited above numerous examples of scholars and individuals who feel indigenous languages are important to the expression of self. Hofman, writing about Zimbabwe, states, "There is considerable awareness that language has an important role to play in whatever national dreams are being dreamt. A national language will help to overcome tribal diversity" (1977, p. 289).

If only language and nationalization issues in many LDCs were as simple as selecting one or a few of the local languages and using them as national languages, to "help overcome tribal diversity," as Hofman mentioned. The example of the Indonesian government's selection of an indigenous language renamed Bahasa Indonesia, as the national language, and actually having it widely accepted within the country, is unfortunately quite unusual. We don't often find such amicable solutions to the raising of an indigenous language to national language status.

What is more common are situations such as the following. Since achieving independence from Britain, India has not been able to unify under the Hindi language, which is seen as giving too much power to Hindi speakers by others who aren't, including Tamil speakers who constitute a large group in the country. Tamil is a Dravidian language (a different family from Indo-European, of which Hindi is a member) spoken widely in Southern India. So we have a north–south split in a country of around 700 million population, plus a very complicated multilingual pattern. Unification is currently still largely expressed through the use of English as an official language.

In Nigeria, the largest country in Africa, Hausa (a Hamito-Semetic language in the same family as Arabic and Hebrew) could be the national language, according to Mazrui (1988), but is resisted by speakers of other Nigerian languages, such as Yoruba or Ibo, which belong to the Sudanese/Niger subfamily of the African family of languages. As a result we again see English, the language of the former colonizers, as the major link language within the country. This pattern is repeated again and again in LDCs, and not just with English.

However, the evident solution of use of the language of the former colonizers is not a happy one for many countries who value sovereignty, and some seem to be using it only as an interim solution while they develop an indigenous national language, such as Shona in Zimbabwe. That does not mean English will not continue as a link language, but probably more on an international rather than an intranational basis.

Does the predominant international use of English in telecommunications technologies pose a problem for those countries valuing sovereignty and defining its attainment as including the use of an indigenous intranational language or languages? Is it a countervailing force which is part of a transnational culture conveyed by the technology? Tsuda (1986) would argue this is the case. In fact, he argues that technology, which includes communications technologies, exerts neocolonial control by the West. Or as Cleveland (1985) puts it, there is less "enclosure" now because of computers linked via global telecommunications. He asserts that power is "leaking" from sovereign national governments. Schiller states the case even more forcefully.

> Thus, communication and the flow of messages and imagry *within* and *among* nations—especially between developed and dominated states—assume a very special significance. What does it matter if a national movement has struggled for years to achieve liberation if that condition, once gained, is undercut by values and aspirations derived from the apparently vanquished dominator? (1976, p. 1)

MODERNIZATION AS A VALUE

On the one hand, then, we find a widespread belief that the use of English does not help achieve sovereignty and decolonization. Yet on the other hand, another major value in LDCs is modernization. Obviously there is a strong desire to benefit from development, to progress, which is usually seen as inextricably tied to modernization.

What is modernization? For one thing, it is products and includes computers, radios, TVs, tape recorders, satellite dishes, and other information, communications technology forms. For another, Dizard defines it as:

> the doctrine of organized universal betterment. As a worldwide civil religion, it is more influential than nationalism or such limited movements as democracy, facism or communism. It shows itself as a psychic mass migration toward a better life. Once this idea makes contact with a society, it diffuses in ways that irrevocably affect traditional institutions and values. It becomes the universal social catalyst, changing everyone it touches. (1982, p. 16)

Or put another way:

> What they all want . . . is what the Americans have got—six lanes of large motor cars streaming powerfully into and out of gleaming cities; neon lights flashing, and juke boxes sounding and skyscrapers rising, story upon story into the sky. Driving at night into the town of Athens, Ohio, four bright colored signs stood out in the darkness—"Gas," "Drugs," "Beauty," "Food." Here, I thought, is the ultimate, the *logos* of our time, presented in sublime simplicity. It was like a vision in which suddenly all the complexity of life is reduced to one single inescapable proposition. These could have shone forth as clearly in Athens, Greece as in Athens, Ohio. They belonged as aptly to Turkestan or Sind or Kamchatka . . . there are, properly speaking, no Communists, no capitalists, no Catholics, no Protestants, no black men, no Asians, no Europeans, no Right, no Left and no Center. . . . There is only a vast and omnipresent longing for Gas, for Beauty, for Drugs and for Food. (Muggeridge, 1978, p. 125)

These two quotations capture much of the essence of the value of modernization so dearly held by so many. It also clearly is felt to rest on technology, and now especially information technology, which is part and parcel of global telecommunications technology. And what language does most of modernization and technological development occur in? Ergo, English *is* the language of development, of technology, of modernization. As nonsensical as that connection is, in the sense that English is in no way inherently better able to express technological thoughts than other languages, it still strongly exists, as I've discussed above. In fact Grabe (1988) asserts that English is not just connected in people's minds to technology, but is actually *necessary* to modernization.

English also does have a "muscular" technological vocabulary already in place and ever growing. That vocabulary forms part of the basis of a common position about the place of English vis-à-vis many indigenous languages:

> Even where indigenous varieties have achieved a developed status they are still not necessarily equal in all sense to external languages. Standardised Guarani and Somali are very much less useful, in a broad perspective, than are Spanish or English, particularly given the desired social mobility and modernisation which now seem to be global phenomena. (Edwards, 1985, p. 85)

This connection will probably become even stronger because of the increasing dominance of telecommunications in helping to achieve modernization. Along with global telecommunications comes the shrinking world, the so-called global village, and what Cleveland calls "the passing of remoteness," which he claims is one of the "great unheralded macrotrends of our extraordinary time" (1985, p. 76). Along with all this comes what others call a "transnational form of culture."

THE ROLE OF ENGLISH AS A LANGUAGE OF WIDER COMMUNICATION

The dominance of English clearly shows the direction of dominance in telecommunications, both hardware and software, so to speak. The First World, and in some cases the Second World, brings its modernizing tools to the Third World, delivering what McPhail (1986) would argue is electronic colonialism, often in the form of this so-called transnational culture. This culture is purveyed by telecommunications, is felt to be a great leveler, but not an equalizer, and is also seen to be a threat to sovereignty, to the rise and development of indigenous languages, and to the preservation of indigenous cultures. The major language of this transnational culture is, of course, English.

Where does this lead for LDCs whose governments and peoples hold the values of sovereignty and modernity? It seems to lead to a double bind, a situation not unlike the fabled Procrustean bed. Or perhaps one prefers the image of the horns of a dilemma. To quote Levi-Strauss, "We are doubtless deluding ourselves with a dream when we think that equality and fraternity will some day reign among human beings without comprising their diversity" (1985, p. 23). Krishnaswamy and Aziz call for just that dream, however, when they assert, "Without losing their identities, nations want harmony; without losing their *valuable values* and cultural heritage, people want better relationships. They want to be Indians/Arabs/Japanese/Chinese etc. and at the same time international" (1983, p. 100; emphasis added).

While we're dealing in images, there's also "having one's cake and eating it too." Edwards argues this is a common reaction to valuing both sovereignty and modernity. "Most people (and most ethnics) want the solace of the past *without* sacrificing the rewards of progress" (1985, p. 42). Gloss *sovereignty, indigenous culture, mother tongue,* and so on for *the past.*

On the other hand, Fishman sees the possibility that nationalism and the value of sovereignty are not quite so much at odds with modernization, but can be very positive. He states:

> Nationalism is not so much backward-oriented . . . as much as it seeks to derive unifying and energizing power from widely held images of the past in order to overcome a quite modern kind of fragmentation and loss of identity. (1972, p. 9)

Given these many serious issues raised by the use of English as a world language, the hegemony of global telecommunications over the world's information structure plus its importance in development, and the role of English in telecommunications, what is its possible future in LDCs which value both sovereignty and modernization? Will it continue to grow as it has in the past, or will it "wither away," as Marx said the state was supposed to do? We know what has happened with the latter; what about the former?

As one would assume, there are those who feel it will continue to spread. Wardhaugh subscribes to this position. "There is no indication that English is in any way ceasing to spread; indeed, it seems to be on the ascendant in the world with no serious competitor" (1987, p. 128). Pretty strong stuff in a world where the United States is the largest debtor nation in existence and where Japan seems to be rapidly moving ahead in economic growth and strength. In southeast Asia, Shaw (1983) asked Indian, Thai, and Singaporean students what they thought the future of English was. A majority of all three groups felt that English would continue to be a world language even if the United States and Britain lost their power. Obviously, its base is now far beyond the borders of countries in which it is a national language, as discussed above. And as the number of nonnative speakers continues to increase over the number of native speakers, it "it increasingly becomes a language belonging to those who use it and not just to those who claim it as their mother tongue" (Shaw, 1983, p. 30).

What factors might contribute to its shrinkage, to its diminished status as an international link language? Will English go the way of French? Or of Latin long before? As indicated above, English now may not be particularly tied to the U.S. and British technological superiority. However, Japan's major gains in this area seem to be encouraging more to learn Japanese, although the high-end estimate of 5 million studying it worldwide is quite small compared to the numbers studying English and some other languages. If Russia and China become world economic powers, which they could well do in the relatively near future, will Russian and Mandarin challenge English? After all, both these countries have populations larger than that of the United States, and Chinese currently has the most native speakers of any language in the world. At this time, we have no way of foreseeing whether or not this is a possibility or probability.

Is it possible that technology, a major reason for the spread of English, can go so far as to contribute to its contractions as a world language? The Japanese are currently working on a simultaneous translator telephone for Japanese and English. In fact, according to Feigenbaum and McCorduck (1984), two experts in artificial intelligence, the Japanese are in "hot pursuit" of natural language processing systems, including creating computers which can understand continuous human speech to 95% accuracy. They feel optimistic that the Japanese have the ability to succeed in this; it may be the case. And Grabe (personal communication, February 27, 1989) indicates he's heard claims for a Chinese-English machine translation system with 80% accuracy.

There are still other factors as well which we should consider, often sociopolitical in nature, such as "neotraditionalization" movements in other parts of the world, where the value of sovereignty is felt to be a way of decolonializing and consolidating (Tsuda, 1986). And if English were to recede as a world language, Fishman suggests relatively "few . . . will shed a tear. The world has no tears left. At any rate, crying takes time and, as all the world has learned from American English, 'time is money'" (1982, p. 21).

In contrast to this tide of attitudes and as a countervailing force, we cannot ignore what seems clear from a great deal of research, namely that "the forces and factors leading to increased *knowledge* of English, *use* of English, and *liking* for English are, nevertheless, usually quite different and unrelated to each other" (Fishman, 1977b, p. 330). In other words, you don't have to *like* English to *use* English. And its use is also increasingly confined to the status Fishman (1977a) calls a "co-language" status, leading to what Verma (1969) called "registral bilingualism," meaning you talk about certain topics to certain people in one language and use another for other topics to other people at other times.

CONCLUSION

Where does this leave us? Does English behave as a window, a mirror, or a kaleidoscope? What can we conclude about its impact on values held by LDCs and about its role as the dominant language used in global telecommunications? As this volume also deals with issues of public interest, perhaps we could rephrase these questions as, "Is the use of English as the dominant language of global telecommunications in the public interest of LDCs who value sovereignty and modernization?"

If English behaves as a kaleidoscope, distorting the message and acting as an instrument of neocolonialism, then it is not in the public interest. But one can validly argue that English, or any other natural language, does not ipso facto distort messages. The receiver may perceive that anyone who uses English distorts the message, or the users themselves may distort, but that is the judgment of the receiver or the decision of the user. Certainly there is a tremendous amount of distortion going on in indigenous languages. One can lie as easily in

Somali, Guarani, Thai, or Hausa as one can in English—and it has been done. There is nothing more "pure" or "pristine" about indigenous languages; they have a different set of attitudes held about them than does English.

What seems to happen with the kaleidoscope image of English is that its users—former colonial "masters" and multinational corporate chiefs, for example—and some of its uses have been mixed up with the language itself, its structure, its vocabulary, and so on. When one holds the position of linguistic relativism, this mixup occurs, Certainly, then, English can be a distorter of the "truth." But what about the uses of German by the Nazis? The perversion of the language in the slogan "Work shall set you free" over the entrance to Auschwitz is not inherent in German but was created by the Nazis, acknowledged masters of propaganda, in a language once considered highly appropriate for science. Or what about the uses of Russian in *Pravda* and other official organs which, I am told, virtually all Russians read between the lines and learn to interpret in very different ways from what is stated in print? I am sure the Khmer Rouge use Cambodian in distorted ways to propound their policy of genocide.

What crimes have been committed in Third World countries by English speakers from Britain and the United States are not *because* of the language. What is important is to separate a language from its users and the uses to which it may be put. We must not succumb to guilt by association, as English can be a powerful tool and in the public interest, as many countries are finding. That is why, in part, the spread of English has been so extensive. It has ways of talking about a variety of technical subjects, as well as literature, art, and the full range of human endeavor.

So is English a kaleidoscope? No, it isn't, inherently, but can be used that way if the user chooses to do so. Then is it a mirror, as I've defined my use of mirror? Does it reflect the West, the values and institutions of the United States and Great Britain, when it's used by a nonnative speaker? I asked Ali Mazrui about this, as he had argued in his book (1975) there may be a Westernizing process implicit in the act of learning English. He now expressed the opinion (personal communication, May 16, 1988) that it was because of the context of language learning, which in Africa is often conducted in Western-style institutions as part of an entire acculturation process. He feels that the values English carries with it in Africa are highly dependent on the context of learning the language and the uses to which it is put in those countries.

Perhaps learning and using a language cannot be completely separated from the culture of its major native speaker groups, as in many cases they do spread together. (This is a highly complex issue which I have not addressed in any detail in this chapter.) But language and culture can be separated far more than they usually are. An example is the English lesson in the Yemeni materials in this chapter. Kachru, in the Public Broadcasting System television series "The Story of English" (1986), stated that he felt most Indians had made English their own and had created an Indian-based English which was not tied in any important way to British English. Other sociolinguistics cited above make the same case. So is English a mirror into which one looks and sees a "brown sahib?" Perhaps, but not necessarily so.

That leaves the window image, the one I feel most describes the possibilities of English or of any other language. This also appears to be the consensus in the field of language planning, which involves many sociolinguists. It may be at times a broken window, or a dirty window, depending on the attitudes of the learners *and* of the teachers, but it *is* a window. It can be a *tool* for the achievement of values such as sovereignty and modernization, and can definitely be in the public interest to use as a link languages at this time in history, as Latin and French were used in times past. In a very real sense, English is a window on the world. It's useful in some domains such as technology, including telecommunications technology, which is involved deeply in modernization efforts. And as Fishman (1977b) pointed out, the data show that liking and usefulness are not connected.

English is being "decolonialized" and even "de-Westernized," in that many local Englishes are appearing throughout the Third World, having their own standards which are not American English or British English. Wardhaugh (1987, p. 15) makes a strong case for this:

> English is the least localized of all the languages of the world today. Spoken almost everywhere in the world to some degree, and tied to no particular social, political, economic, or religious system, nor to a specific racial or cultural group, English belongs to everyone or to no one, or it at least is quite often regarded as having this property.

In sum, does English, which is intimately tied to technology and conveyed world-wide by telecommunications technology, support or not an LDC's moves toward sovereignty and modernization? In terms of the value of sovereignty, each country will have to find its own way, so to speak, among the language choices available to it. However, as English can be a window, a relatively neutral vehicle for communicating nonneutral uses, it does not necessarily have to be discarded in the pursuit of sovereignty. Yet in many LDCs, it does have diminished use as an intranational language for expressions of sovereignty.

In terms of the value of modernity, English's role is more clear in the domains of information, technology, and science and their relationship to development in LDCs. English provides a window to a some of the know-how for creating a better quality of life for the majority of the world's people. This is not to say that the use of English will automatically provide access to that knowledge; e.g., it is not a sufficient condition for development, but in many cases it is a necessary condition.

Let me end this chapter and this summary with a quote from Kaplan, who wrote in an editorial in *Science:*

It seems reasonable to assert, however difficult it may be to accept, that knowledge
of a world language, especially English, is essential to the welfare of the new
nations . . . New nations must find a balance between the cultivation of indigenous culture-rich language and the need for a world language . . . Any other
course is tantamount to restricting their capability for modernization. (1983, p.
4614)

Note he says "however difficult it may be to accept," indicating that charges
of linguistic and cultural imperialism may be leveled at writers such as myself. I
am not propounding that English *should* be the dominant link language in the
world today and continue to spread at its present rate; I am simply indicating
that it *is*. This may or may not be a desirable state of affairs, but if LDCs wish to
modernize while asserting their sovereignty, they can at least do so with an
understanding of language as a window, not just a mirror or kaleidoscope.

REFERENCES

Baxter, J. (1983). Interactive listening. In L. Smith (Ed.), *Readings in English as an
international language* (pp. 103–110). Oxford, England: Pergamon Press.

Cleveland, H. (1985). The twilight of hierarchy. In B. R. Guile (Ed.), *Information
technologies and social transformation* (pp. 55–79). Washington, DC: National
Academy Press.

Cohen, M. (1988, June 8). Malaysian students struggle to cope with language shift.
Chronicle of Higher Education, pp. A29–30.

DeStefano, J. (1986). Commentary on session, New communications technology and
international academic exchange. In A. R. Devereux & G. L. Seay (Eds.), *Minds
without borders: Educational and cultural exchange in the twenty-first century,* The
Fulbright Fortieth Anniversary Washington Conference Proceedings (pp. 118–119).

Dizard, W., Jr. (1982). *The coming information age.* New York: Longman.

Edwards, J. (1985). *Language, society and identity.* New York: Basil Blackwell.

ETS Developments, 33 (3, 4), Winter/Spring 1988.

Feigenbaum, E. A., & McCorduck, P. (1984). *The fifth generation.* New York: Signet.

Fishman, J. (1972). *Language and nationalism.* Rowley, MA: Newbury House.

Fishman, J. (1977a). English in the context of international societal bilingualism. In J.
Fishman, R. Cooper, & A. Conrad (Eds.), *The spread of English* (pp. 329–336).
Rowley, MA: Newbury House.

Fishman, J. (1977b). Knowing, using and liking English as an additional language. In J.
Fishman, R. Cooper, & A. Conrad (Eds.), *The spread of English* (pp. 302–310).
Rowley, MA: Newbury House.

Fishman, J. (1977c). The spread of English as a new perspective for the study of "language maintenance and language shift". In J. Fishman, R. Cooper, & A. Conrad
(Eds.), *The spread of English* (pp. 108–133). Rowley, MA: Newbury House.

Fishman, J. (1982). Sociology of English as an additional language. In B. Kachru (Ed.),
The other tongue, English across cultures (pp. 15–22). Urbana, IL: University of
Illinois Press.

Fishman, J., Cooper, R., & Conrad, A. (1977). *The spread of English*. Rowley, MA: Newbury House.

Grabe, W. (1988). English, information access, and technology transfer: A rationale for English as an international language. *World Englishes, 7*(1), 63–72.

Hofman, J. (1977). Language attitudes in Rhodesia. In J. Fishman, R. Cooper, & A. Conrad (Eds.), *The spread of English* (pp. 277–301). Rowley, MA: Newbury House.

Intermedia. (1973). *1*(3), p. 1.

International Commission for the study of Communication Problems. (1980). *Many voices, one world* (McBride Commission Rep.). Paris: UNESCO.

Kachru, B. (1976). Models of English for the third world: White man's linguistic burden or language pragmatics? *TESOL Quarterly, 10*(2), pp. 221–239.

Kachru, B. (1982a). Introduction: The other side of English. In B. Kachru (Ed.), *The other tongue, English across cultures* (pp. 1–12). Urbana, IL: University of Illinois Press.

Kachru, B. (1982b). Models for non-native Englishes. In B. Kachru (Ed.), *The other tongue, English across cultures* (pp. 31–57). Urbana, IL: University of Illinois Press.

Kachru, B. (1983). *The Indianization of English: The English language in India*. Delhi, India: Oxford University Press.

Kaplan, R. B. (1982, May). *Information science and ESP*. Paper presented at the 16th Annual TESOL Convention, Honolulu, HI.

Kaplan, R. B. (1983). Language and science policies of new nations (Editorial). *Science, 221,* 4614.

Kaplan, R. B. (1987, March). *Language policy in the Pacific Rim*. Paper presented at the 1987 TESL Canada Conference, Vancouver, BC.

Krishnaswamy, N., & Aziz, S. (1983). Understanding values, TEIL and the third world. In L. Smith (Ed.), *Readings in English as an international language* (pp. 95–102). Oxford, England: Pergamon.

Laitin, D. (1977). *Politics, language and thought: The Somali experience*. Chicago, IL: University of Chicago Press.

Levi-Strauss, C. (1985). Race and cultures. In C. Levi-Strauss (Ed.), *The view from afar* (pp. 3–24). New York: Basic Books.

MacNeil, R. (Producer). (1986). *The Story of English* [television series]. Washington, DC: The Public Broadcasting System.

Mandelbaum, D. (Ed.). (1956). *Edward Sapir: Culture, language and personality: Selected essays*. Berkeley, CA: University of California Press.

Mazrui, A. (1975). *The political sociology of the English language: An African perspective*. The Hague, Netherlands: Mouton.

McCrum, R., Cran, W., & MacNeil, R. (1986). *The story of English*. New York: Viking.

McPhail, T. (1986). *Electronic colonialism: The future of international broadcasting and communication* (Rev. 2nd ed.). Newbury Park, CA: Sage.

Muggeridge, M. (1978). *Things past*. London: Collins.

Schiller, H. (1976). *Communications and cultural domination*. White Plains, NY: International Arts and Sciences Press.

Shaw, W. (1983). Asian student attitudes toward English. In L. Smith (Ed.), *Readings in English as an international language* (pp. 21–34). Oxford, England: Pergamon.

Strevens, P. (1987). English as an international language. *English Teaching Forum, 25*(4), 56–64.

Tonkin, H. (1984). A right to international communication? In G. Gerbner & M. Siefert (Eds.), *World communications, A handbook* (pp. 69–79). New York: NY: Longman.

Tsuda, Y. (1986). *Language inequality and distortion in intercultural communications, a critical theory approach.* Philadelphia, PA: John Benjamins.

Verma, S. K. (1969). Towards a linguistic analysis of registral features. *Acta Linguistics Academiae Scientarum Hungaricae, 19,* 293–303.

Wardhaugh, R. (1987). *Languages in competition, dominance, diversity, and decline.* New York: Basil Blackwell.

Weinstein, B. (1983). *The civic tongue: Political consequences of language choices.* New York: Longman.

Whorf, B. (1956). Linguistics as an exact science. In J. Carroll (Ed.), *Language, thought and reality: Selected writings of Benjamin Lee Whorf.* Cambridge, MA: MIT Press.

chapter 5
Democracy, Technology, and Privacy*

Sven B. Lundstedt
Ameritech Research Professor
Professor of Public Policy and Management, and International Business
School of Public Policy and Management and Faculty of Human
    Resources and Management
The Ohio State University

In 1974, the late Senator Sam J. Ervin, Jr., of North Carolina, then Chairman of the U.S. Senate Subcommittee on Constitutional Rights, wrote:

> A government called upon to manage an increasingly complex modern society and to satisfy ever-widening demands of the people for services has come to require more and more information. . . . Only in the last few years has it become widely recognized that the new information technology gives government great opportunities to do ill, as well a good.
>
> The Founding Fathers knew well that with power comes the ability to do harm. The fundamentals of our constitutional system require us always to ensure that governmental power is sufficiently constrained by law so that much as is humanly possible the power of government is used for good alone, and that our nation continues to have a government subject to the people, and not the reverse. We have slowly come to the realization that this is true no less for information practices as it is for other Government activities. (Staff of the Subcommittee, 1974)

The subcommittee's early investigations of government data banks and individual rights revealed not only a disturbing absence of laws to control the new information technologies, but also a lack of knowledge of what data banks the government possessed, what was in them, and how they were used.

Several discoveries illustrate the senator's concerns. The first was a Secret Service memorandum requesting information about persons who made anti-government statements or embarrassing comments about government officials. The second was a disclosure that the Department of Health, Education, and Welfare had blacklisted prominent scientists from being appointed to advisory

---

boards because of political views. A third was a revelation that the Army had devised a computer system to conduct political surveillance.

Following the subcommittee's survey, Senator Ervin said investigators "discovered numerous instances of agencies starting . . . with a worthy purpose, but . . . going so far beyond what was needed in the way of information that the individual's privacy and right to due process of law are threatened by the very existence of the files." The most significant subcommittee finding was that large numbers of data banks existed that were full of diverse information "on just about every citizen in the country." Indeed, 54 federal agencies reported 858 data banks containing more than 1.5 billion records on American citizens.

We may only assume that these records have increased in the past 15 years, thus making the constitutional issues raised even more compelling today. Indeed, the growing presence of an unchecked information technology poses a serious threat to individual freedom.

EXAMPLES OF SURVEILLANCE TECHNOLOGY

The range of communication technology that can be used for electronic surveillance is growing constantly. Electronic eavesdropping technology for both audio and visual surveillance includes miniaturized transmitters; wire systems involving telephone taps, concealed microphones, and tape recorders; optical/imaging technology, such as photographic techniques; closed circuit and cable television; and vision devices using image intensifiers to see objects in the dark.

There are computers and related technologies for data surveillance utilizing networking, expert systems software, and pattern recognition systems. There are also sensor technologies such as magnetic sensors, infrared sensors, strain sensors, and electromagnetic sensors. Yet another group is comprised of citizen band radios, vehicle location systems, machine-readable magnetic tapes, polygraphs, voice stress and voice recognition analyzers, laser reception, and cellular radios.

These devices can detect and measure a wide range of human activity: movements of people in time, actions of people, such as the number of keystrokes they make on computer terminals; access to financial and commercial accounts records; computerized law enforcement and investigatory records; and verbal communications that take place on telephones or electronic devices. Electronic visual surveillance also can monitor details of behavior in public or private places, day or night. Perhaps the most obtrusive of all the new information technologies are those that measure internal behavior, including polygraphs, voice stress analysis, breath analyzers, and brain wave analyzers.

Use of this sophisticated technology for surveillance carries obvious dangers in a democratic society. However, the study of human behavior in legitimate scientific research also raises the issue of invasion of privacy. This is a complicated legal, ethical, and philosophical problem that has large been managed (or mismanaged) by government regulation.[1]

THE NEED FOR PRIVACY SAFEGUARDS

The 1974 Senate subcommittee study demonstrated a need for statutory authority to re-examine pre-existing data banks and legislative approval to create new data banks. It also showed that privacy safeguards should be built into computerized government files as they are developed. People should be told that information about them is being stored in a federal data bank. And they should have the right to see and correct that information. Interagency exchanges of personal data should be constrained and interagency data bank cooperatives established. There should be strict security precautions to prevent unauthorized or illegal access to data banks. Finally, there is need for continuing legislative control over the purposes, contents, and uses of new information technologies. Many of these controls have not yet been created; however, the technology—for good and ill—is not in place.

The 1972 National Academy of Sciences' (NAS) *Databanks in a Free Society*, based on site visits to 55 organizations, found that computers and data banks have not led inevitably to greater collection and data manipulation. Organizations seem to have adhered to traditional democratic principles concerning data collection and sharing (Westin & Baker, 1972; for additional information on this subject see Linowes, 1989).

The study concluded, however, that sensitive information remained largely in manual form. It said that an impending explosion of computer information technology was about to increase dramatically the number of information networks —a trend with potential negative impacts upon individual privacy. As information technologies continue to increase, so have information networks, along with an increased potential for violations of individual rights to privacy.

William Fielding Ogburn's (1957) insightful concept of "cultural lag" brings to light a dilemma. If the pace of information technology is largely exponential and if human behavior and institutions are slow to respond, we cannot expect Congress, public law, and the courts to keep up. If this is true, can the new public policies necessary to deal with rapid technological change be formulated in time?

The potential for mischief increases daily if we add to the applications of information technology in government those taking place in the private sector in telecommunications, banks, credit bureaus, company personnel files, and hospital records.

---

[1] For examples of relevant Supreme Court rulings concerning use of electronic surveillance, see Office of Technology Assessment (1987, pp. 24–25).

As NAS noted, most of this information is accessed and used by well-meaning bureaucrats. However, we can add that it also provides "hackers" and other renegades—as well as public officials motivated by personal or political beliefs—an opportunity to misuse the information on an unprecedented scale.

Paradoxically, legislation to protect the individual's right to know—the Freedom of Information Act, which became law in 1966—also created new opportunities for unwarranted invasions of privacy by opening government files to the public. Consequently, the 92nd Congress in 1971 explored several proposals to amend the Freedom of Information Act to increase the protection of individual privacy. An example is legislation that would have prohibited federal agencies from distributing lists of names and addresses of individuals and the creation of a federal privacy board. Many proposals were based on the principles that shaped the Fair Credit Reporting Act, which was enacted by the 91st Congress to limit abuses in credit reporting.

Information systems used by criminal justice and law enforcement agencies, including the FBI, constitute a major arena for potential privacy rights abuse. Unwise, unethical, and politically motivated use of information systems at the federal level also sets a poor example for local and state law enforcement agencies.

There are legitimate reasons and places for secrecy—in diplomacy and international relations, the business world, intelligence work, and law enforcement. For the most part, such concerns for secrecy are not at odds with the Constitution. What is at issue are those secret acts that depart from the norm set by the Constitution. The debate is often intellectually difficult, however, because the concept of privacy is not clearly defined either philosophically, legally, or practically.

RECENT PUBLIC POLICIES AND EXAMPLES

In 1986, Congress updated a 1968 law to include the protection of privacy for new communication technologies such as cellular telephones, paging devices, electronic mail, and private satellite transmissions. This legislation was designed to enhance protection against illegal eavesdropping, including new forms of "wiretapping" that involved electromagnetic, photoelectric, or photo-optical systems. It also became illegal to intercept electronic communication without a court order, a procedure designed to protect the privacy of stored and transferrable communication.

Congress also has moved to control the potentially harmful effects of computer matching, which could lead to invasions of privacy. A pending House bill would require federal agencies to enter into computer-matching agreements only after the purposes of comparing the data, the nature of the match, and the expected results have been specified. Individuals must be given notice and must have a chance to contest findings derived from matching. The Senate has passed a similar bill.

One final example of recent activity is illustrated by the concerns expressed by Congressman Don Edwards of California, who requested a special report on the new FBI initiative to expand its data network as part of its National Crime Information Center. The proposed system includes tracking suspects not yet charged with any crime. The report, prepared by the group Computer Professionals on Social Responsibility, was a basis for hearings held in March 1987.

Congressman Edwards serves on the House Subcommittee on Civil and Constitutional Rights, which has oversight jurisdiction over the FBI. The subcommittee has an interest in the civil rights implications of the use of computers in law enforcement. The concerns that led to the report included the constitutional dangers associated with a proposal to establish files on people who were merely suspected of criminal activity but not charged. This proposal undermines the presumption of innocence, a fundamental Constitutional principle. As in the earlier Ervin committee study, there is also renewed concern about the privacy violations that could occur because of networking linkages with other government systems. A third issue is that of maintaining the reliability and validity of data in files. This is directed, for example, at reducing the number of mistaken identities that may result in damaged reputations and wrongful convictions (Horning, Neuman, Redell, Goldman, & Gordon, 1989).

These new bills illustrate the form of legislative activity that followed the Senate subcommittee study chaired by Senator Ervin. In fact, the computer-matching legislation may be seen as the fulfillment of a recommendation made by the Ervin subcommittee rather than a new measure. However slow and uncertain, incremental legislation, nevertheless, conforms to U.S. political traditions that honor the practices of representative government, openness, and debate.

The National Telecommunications and Information Administration recently issued a comprehensive review of the communication and information sectors that highlighted a continuing need for Fourth and Fifth Amendment privacy safeguards and identified them as a national priority (U.S. Department of Commerce, 1988). The report singled out interception of communications, government and private sector recordkeeping, and security of information as special areas of concern. A recent example is a new electronic device that can be used by telephone subscribers to display incoming numbers before answering the phone. (Sims, 1989) This form of technology intervention has several unwanted privacy side effects, such as disclosing an unlisted number without the caller's consent and the use of such numbers by others that can lead to a breach of consent. Technology may offer a solution. Pacific Telesis plans to introduce a form of encryption.

A recent study by the Officer of Technology Assessment (1987) discusses technological issues (both software and hardware) associated with problems of information security and privacy. In a chapter about the vulnerabilities of electronic information systems, the study concludes that telephone networks continue to provide opportunities for targeted and untargeted eavesdropping. As in the example of Pacific Telesis, security and safeguard technologies using encryption are also increasing. However, there has always been a tendency to rely too much upon "technological fixes." Therefore, institutional and organizational standards—both legal and ethical—also need to be developed.

INDIVIDUAL INVIOLABILITY

Part of the intellectual dilemma concerning privacy issues can be attributed to the fact that we do not have an adequate definition to use in particular cases. It is a complex idea, like the idea of the public interest, for which there are numerous definitions, most of which are inadequate and hard to apply. Consequently, the court and legislature are in need of a clearly formulated and evolving working concept that can be used in particular cases. Fortunately, in the history of Western thought, there are numerous examples, in words and actions, that emphasize individual inviolability. These can help move us toward such a definition. Without important consideration of the idea of privacy, we may lose the benefit of a deeper intellectual analysis and continue to substitute legal precedents and analysis for it.

We owe a debt to John Stuart Mill, among others, for the following clarification of the idea of privacy—a clarification made more than a century before new information technologies complicated the issue:

> The appropriate region of human liberty . . . comprises first, the inward domain of consciousness; demanding liberty of conscience in the most comprehensive sense; liberty of thought and feeling; absolute freedom of opinion and sentiment on all subjects, practical or speculative, scientific, moral, or theological. . . . Secondly, the principal requires liberty of tastes and pursuits; of framing the plan of our life to suit our own character; of doing as we like, subject to such consequences as may follow; without impediment from our fellow creatures, so long as what we do does not harm them, even though they should think our conduct foolish, perverse, or wrong. (Mill, 1952, pp. 272–273)

Thus the indifferent force of a communication technology requires a value framework that (a) defines how the technology should be used morally and politically, and which (b) signals when the technology becomes a danger to individuals, communities, and society. The exploitive, sometimes malevolent, uses of political and police power—when joined with the power of technology—have in the past resulted in extraordinary tragedies. With new information technologies, there is even a greater need for intellectual clarity and vigilance by the Congress, courts, and executive branch to assure that wrong-doings do not occur.

OPENNESS, SECRECY, AND POLICY STYLES

In all policy activity there is an important distinction between "latent policy" (one held or aspired to covertly) and "manifest policy" (where all cards are face up on the table).[2] The conflict between latent policy and manifest policy lies at the heart of many of the problems of liberty, privacy, and information technology. Furthermore, if latent policy is malevolent, even if dressed in benevolent clothing, the situation is even more dangerous—as Watergate and the Iran-Contra affairs have illustrated. Each led to an abandonment—often with the aid of sophisticated but secretive technology—of that fundamentally essential institution of civilized societies: rule by law.

Such issues and violations are as old as the Republic itself. However, the exponential growth of information technology and its proper management will make these issues a preeminent challenge for democratic societies for years to come.

REFERENCES

Horning, J. J. et al. (1989). A review of NCIC 2000: The proposed design for the National Crime Information Center. Report submitted to the Subcommittee on Civil and Constitutional Rights of the Committee on the Judiciary, United States House of Representatives.

Linowes, D. F. (1989). Privacy in America: Is your private life in the public eye? Urbana: University of Illinois Press.

Lundstedt, S. B. (1976). Latent policy. The Ohio State University College of Administrative Science Working Paper Series. (Reissued in WPS 81-81, 1981.)

Lundsted, S. B., & Spicer, M. W. (1977). Latent policy and the Federal Communications Commission. The Ohio State University College of Administrative Science Working Paper Series (WPS 77-12).

Mill, J. S. (1952). On liberty. In The great books of the Western World (pp. 272–273). Chicago: Encyclopedia Britannica, Inc.

Ogburn, W. F. (1957, January). Culture lag as theory. Sociology and Social Research, XLI, 167–173.

Office of Technology Assessment. (1987). Defending secrets, sharing data—New locks and

---

[2] "Latent Policy means an intentionally or unintentionally concealed policy . . . The distinction between manifest and latent directs attention toward the unanticipated consequences of policies which lie beyond, and behind, the usual formal procedures in organizations" (Lundstedt, 1976, 1981, pp. 4, 5). For an application, see Lundstedt and Spicer (1977).

keys for electronic information (OAT-CIT-310). Washington, DC: Government Printing Office.

Sims, C. (1989, March 1). Who's phoning? New system will tell you. *New York Times*, p. 1.

U.S. Congress, Senate Subcommittee on Constitutional Rights of the Committee on the Judiciary. (1974). *Federal data banks and Constitutional rights: A study of data systems on individuals maintained by agencies of the United States Government* (93rd Congress, Part III of the Subcommittee's Study of Federal Data Banks, Computers and the Bill of Rights, III, pp. ix–xlvii). Washington, DC: Government Printing Office.

U.S. Department of Commerce. (1988, October). *NTIA Telecom 2000—Charting the course for a new century* (Special Publication 88-21). Washington, DC: Government Printing Office.

Westin, A. F., & Baker, M. A. (1972). *Databanks in a free society: Computers, record keeping and privacy.* New York: Quadrangle Books.

chapter 6
Coping with Telepower

Joseph N. Pelton
Professor and Director
Interdisciplinary Telecommunications Program
University of Colorado, Boulder

In the 14th century, Peking (Beijing) was the world's largest city and, as such,
some 50 square kilometers in area. Thus, one could reach the opposite
than 5 miles. The transportation systems of
hundreds of

Let u
where we a
population ha
human informati
(or computerized) te
to 10^{15} bits of data. In s
human information was ex
base has grown 200,000 times
fast Ferrari. This means it is a
renaissance person today than it was
keep up with state of the art research
man running to catch a space ship. It m
success is very close to zero.

Figure 6.7 shows another way our society has
the cost of a robotics device operation in industry
graph indicates that robotics are already remarkably
U.S. workers, in terms of cost per hour. Perhaps more su
shows that, in the foreseeable future, the same results will a
called "labor-cheap" countries such as Taiwan, the Republic of K
Brazil, and the like.

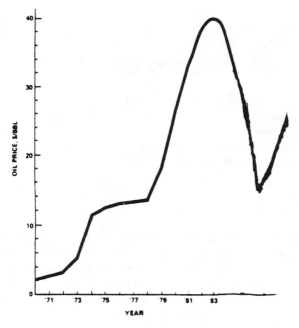

Figure 6.1. Cost Trends for Petroleum

s now look at information and communication services both in terms of
re and where we are going. Since the time of ancient Greece global
s increased dramatically, on the order of 50 times. Collective
on or storable data has increased considerably more. In digital
rms the total human knowledge base has increased from 10^8
hort, while human beings increased in number 50 times,
panding 10 million times. Our collective knowledge
faster. It is like a race between a slow snail and a
out 10 million times more difficult to be a
in the good old days. A generalist trying to
n a field like biology or physics is like a
y be fun to try, but your chance of

changed forever. It compares
ersus the cost of labor. This
nore cost effective than
rprisingly, this graph
lso be true for so-
rea, Malaysia,

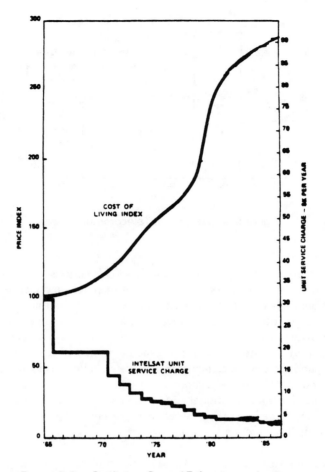

Figure 6.2.   Declining Cost of Telecommunications

What then do all these facts and figures mean, when considered as a whole? Let's reflect upon some key facts:

1. The cost of automation and intelligent robotic industrial production is dropping;
2. The importance of telecommunications and information services is increasing, as reflected in both investment and employment levels;
3. The cost of information services and telecommunications is also dropping in contrast to most other services and products where the cost is being driven upward with inflationary cost spirals. Wages, however, are increasing in

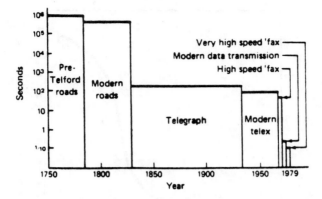

Figure 6.3   Progress in Telecommunications

both developed and developing countries. This contrast is particularly no-
ticeable when comparisons are made between telepower enterprises and
traditional labor-intensive activities. The net result is that jobs are being
lost, with robotics, telecommunications, and information services all serving to reduce industrial jobs and soon service jobs as well; and

4.   The patterns of international investment, the rise of multinational enter-
     prises, and the cost efficiency of worldwide global networks all indicate the
     development of a global economic enterprise based upon geographic de-
     centralization and functional distribution of effort across national lines,
     even though the hub of centralized management and policy control may be
     located halfway around the world.

In short, the world of global talk and global think is arriving in a big way, at
least for a significant proportion of the world's population. The current and
perhaps growing gap between those who inhabit the world of the electronic
future and those who do not could well become a dangerous one. In the 1950s
the developing countries, with two-thirds of the world's population, had over
30% of the world trade; today that proportion has shrunk to 17%. Should this
trend continue, the problem of an "information and communication gap," as
well as the trade gap, could become increasingly troublesome.

If current patterns of technical development and global trade continue, we
can expect to see

1.   The continued rapid emergence of global computer-communications net-
     works plus greatly expanded use of robotics (these two factors constitute key
     drivers of today's world economic activity).
2.   These trends will give rise in technological unemployment and skill loss in

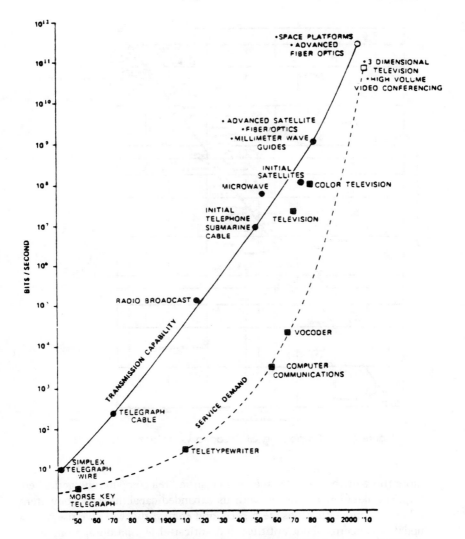

Figure 6.4.   150 Year Look at Development of Telecommunications (Service Demand vs. Transmission Capability)

an increasing number of countries. Legislative reactions almost inevitably will be required in terms of retraining, limits on off-shore electronic services, etc.

3. We will begin in the 1990s to see long-term conversion from conventional petrochemical energy sources to technology-intensive recyclable energy sources or "smart" production. This conversion should also serve to pro-

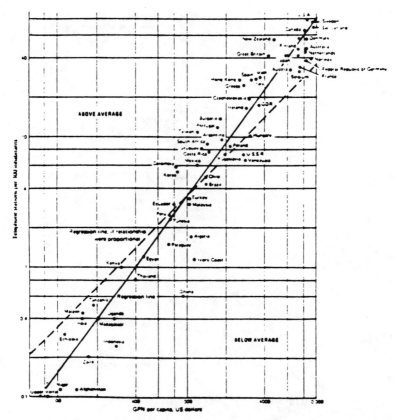

Figure 6.5.   Comparing GNP/Capita vs. Telephones/Capita

mote the growth of newly emerging gigantic "telecomputerenergetics" enterprises like General Motors with its expanded capabilities in computers,
robotics, communications, aerospace, and energy; this will give rise to new
updated concerns about antitrust and anticartel legislation appropriate to
the 1990s, when there is a concern not only about "price financing, but also
technology fixing."

4.   We will also likely see the growth of technology enclaves around resource-
      rich areas to become mega-"silicon valleys" (perhaps we will even see the
      emergence of "floating island" enclaves operated by multinational enter-
      prises).[1] This would be the ultimate offshore activity, at least until we have
      competitively run space stations or space colonies.

---

[1] For further information on the above projections, see Pelton (1981). Also see "Artificial Islands"
(1981).

Agricultural Activity as Percentage of Total GNP	Corresponding Telephone Density
70%	2% or Less
50%	10% or Less
Less than 2%	Up to 70%

Figure 6.6

Current technology trends suggest that the entire fabric and structure of society (in terms of how we work, live, and relax) is changing. The design and functions of our cities, the decisions as to where industry is located, the process by which we manufacture products, the way we grow food, how services are offered and distributed (the very nature of our lifestyles) will all be dramatically affected by what might be called our "future electronic environment".

But the future electronic environment does not necessarily apply to everyone. It is far from clear the extent to which modern electronic communications and energy technology will impact upon the standard of living and societal paths for developing, in contrast to developed, countries. Figure 6.8 shows the startling contrast that exists between the Organization for Economic Cooperation and Development (OECD) member countries and the rest of the world. If we were to assume that the "global village" consisted of 1,000 people, 125 would live in the OECD countries. We would find that, of the remaining 875 villagers,

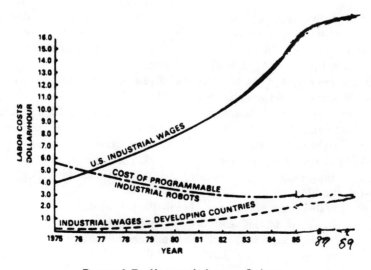

Figure 6.7   Human Labor vs. Robots

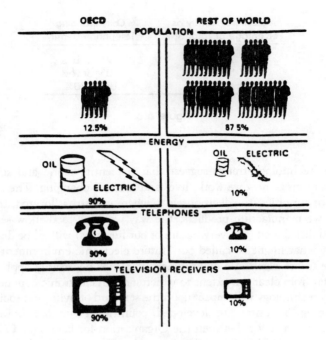

Figure 6.8.   Global Village

over 500 would suffer from malnutrition. The "other" 875 villagers would also have access to only 15 telephones and a handful of mainframe computers. The cost of upgrading telephone service for the remainder of the global village in order that it would be comparable to OECD standards (at $2,000 per installed telephone line) would be a staggering $8 trillion, or equivalent to about half the gross national product of the world for a year.

Clearly, enormous shifts to economic investments would have to occur if indeed the entire world were to become a global electronic village. There is, of course, an important subsidiary question: What are the economic, social, and political implications if the new electronic environment being created for the affluent "first world" is largely isolated from the trends toward economic and technical progress in the so-called Third World countries? Let us focus, for the purpose of illustration, on the pace of change for the world at large, as contrasted to the U.S.

Let us briefly review how things were in 1964, how they are today, and how we might find them in the year 2000. Figure 6.9 shows us a picture of quite remarkable change, especially in the field of communications. Over the 36-year period international communications capabilities will have increased a thousand-fold.

THE U.S. AND THE WORLD: IN THE YEAR 1964, 1987, AND 2000

Indicator	1964	1987	2000 (Projected)
Population (World)	3,100 Million	5,000 Million	6,000 Million
Population (US)	190 Million (6.0%)	240 Million (5%)	290 Million (4.8%)
Electrical Power (World)	3,000 Billion KwH	10,000 Billion KwH	24,000 Billion KwH
Electrical Power (US)	1,100 (36.1%) Billion KwH	2,600 (26%) Billion KwH	4,600 (19.1%) Billion KwH
Telephones (World)	175 Million	600 Million	1,630 Million
Telephones (US)	80 (45.7%) Million	225 (37.5%) Million	422 (26.7%) Million
World Transoceanic Telephony Service*	678 Voice Circuits	80,000 Voice Circuits	Equivalent of 2,000,000 Voice Circuits
US Transoceanic Telephony Service*	330 (48.7%) Voice Circuits	20,000 (25%) Voice Circuits	175,000 (21.8%) Voice Circuits

Figure 6.9.   The U.S. and the World in the Year 1964, 1987, and 2000

In 1964, the highest-capacity transoceanic communications facilities were the TAT-3 cable across the Atlantic and the TRANSPAC cable across the Pacific Ocean. Each had a total capacity of only 138 3-kHz telephone circuits and indeed had only been in operation a few years. It is surprising to some that the first Atlantic telephone cable was laid as recently as 1956 (about 35 years ago).

The era of commercial satellite communications began in 1965. In that year the initial U.S. defense communications satellite system and the Soviet Molniya domestic communications satellite system were both deployed in medium-orbit operation. More significantly, however, the first INTELSAT satellite (Early Bird) was launched in April of that year. Early Bird doubled telephone capacity between the U.S. and Europe and, for the first time, enabled live television transmissions to occur. Thus, 1965 marked the birth of the age of global television. Not as well understood is the fact that "the age of supertribalization" also began with age of satellites. Certainly, the rapid flow of electronic information at low cost and with high reliability has served to create a more unified and uniform world. Cultural differences, language barriers, national marketing strategies, and almost everything else are today enormously influenced by electronic global communications.

Through the global satellite communications system, global TV has become a reality. Five hundred million people watched the moon landing in 1969, and a billion people saw at least some part of the 1976 Summer Olympics in Montreal. In 1984, perhaps 2 billion people saw the Los Angeles Olympics. Over 2 billion saw the Seoul Summer Olympics and the Political Conventions in 1988. Similar dramatic shifts have occurred in the field of computers during the same time period. Growth in the memory capability of computer disk storage devices has averaged over 50% per annum, growing from 5.4 trillion terabytes of memory in 1975 to 70 trillion terabytes in 1982 and over 500 trillion terabytes today. The U.S. computer industry is manufacturing sufficient quantities of processors, memory, and logic circuits to be able to produce more than one mainframe computer for every other American. Obviously, this is not going to be the case. Instead, we will see dozens of microprocessors hidden inside mass consumer goods, such as television sets, toys, automobiles, household appliances, security systems, etc. Recently, an enterprising British group even announced the invention of an "intelligent brassier." This is a personal body temperature monitoring device that aids in the use of the rhythm method of birth control. I'll not explain further how it operates.

The principal focus for our purposes, however, should be on host mainframe devices tied into communications networks that can support a large number of mini- and microcomputers. Figure 6.10 shows the rapid growth of mainframe computers and computer chip capacities, even on a logarithmic scale.

The combination of computer and advanced communications networks today represents the leading edge of technology and, as such is seen by many as a threat and an assault on traditional values. Discussion of this topic therefore must consider the subjective as well as objective aspects of field. Tariff policies, copyright protection, patent and data rights, technical standards, trade agreements, etc. (all of these subjects and more), as they relate to computer communications networks, must be seen in a broader aspect. The issue of whether or not there should be a 7- or 8-bit companding law, or whether TDMA systems should be plesiosynchronous or not, or what the exact ISDN standards should be, all seems arcane, even trivial. Yet these decisions on information standards can also mean millions of dollars of profit or loss. Commercial advantage, market protection, regional interest, and more are frequently involved in "technical decisions." An attempt to broadly categorize some of the key conflicting values in a technological world is provided in Figure 6.11.

If the pattern of change has seemed rapid in the past, the future prospects seem even more challenging. To appreciate the nature of our changing world, let us look at the dimensions of the world of global talk, as measured by digital communications or bits per second. Thirty bits per second may be considered equivalent to one word per second. Most people, therefore, talk at a rate of

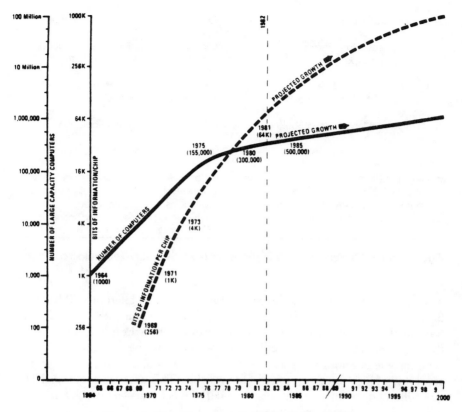

Figure 6.10.   Computer Trends: 1964—2000

around 100 bits per second, when measured in terms of equivalent written text. These slow rates are infinitesimal when compared to our new telepower tools.

Figures 6.12 and 6.13 show us the changing dimensions of our "electronic world." An advanced communications satellite or fiber optic system can today send 1 to 3 billion bits of information a second; today's fastest computers can also process data at these rates as well. Sending data at a gigabit/second (which is 50 times the capability of 20 years ago), it should be noted, is the equivalent of transmitting the complete Encyclopedia Britannica plus all its color graphics at six times a minute.

But these are just the early stages of digital communications revolution. In another 20 years, today's capabilities will seem like a snail's pace, as data transmission rates of perhaps 100 gigabits per second are realized. But let us scale these data rates to human dimensions. A typical fairly literate person processes about 650,000,000 words, or about 20 billion bits of information in a lifetime. If

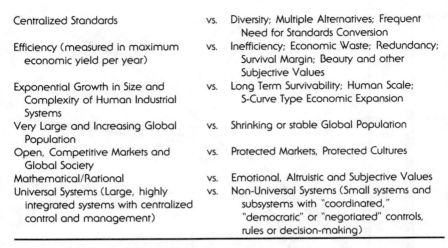

Centralized Standards	vs.	Diversity; Multiple Alternatives; Frequent Need for Standards Conversion
Efficiency (measured in maximum economic yield per year)	vs.	Inefficiency; Economic Waste; Redundancy; Survival Margin; Beauty and other Subjective Values
Exponential Growth in Size and Complexity of Human Industrial Systems	vs.	Long Term Survivability; Human Scale; S-Curve Type Economic Expansion
Very Large and Increasing Global Population	vs.	Shrinking or stable Global Population
Open, Competitive Markets and Global Society	vs.	Protected Markets, Protected Cultures
Mathematical/Rational	vs.	Emotional, Altruistic and Subjective Values
Universal Systems (Large, highly integrated systems with centralized control and management)	vs.	Non-Universal Systems (Small systems and subsystems with "coordinated," "democratic" or "negotiated" controls, rules or decision-making)

Figure 6.11.   Opposing Value Systems for a 21st Century Technological World

we equate 20 billion bits to a Typical Information Use Per Lifetime (TIUPIL), we find that an INTELSAT VI satellite scheduled to be launched in 1989 could transmit 9 TIUPILs in a minute. Going a step further into the future, let us assume that a global network of 10 satellites were available in the year 2000, that each were capable of transmitting at 100 gigabits per second, and that thus, a system capability of a terabit per second were available. This network could send 31.5 quintillion bits of information, or 2.62 GHIUDs per year—a GHIUD, incidentally, is Global Human Information Use Per Decade.

Other than the obvious answer that these calculations are somewhere between far-fetched and mad, what does this suggest? My own answer is, first, that communications in the year 2000 will be frequently of a scale and magnitude that humans cannot assimilate. Most communications will be largely machine-to-machine—perhaps more than 90% or so. Specifically, you will see computer-to-computer communications, or even artificial intelligence-to-artificial intelligence, as the primary means of information relay and storage. The recent stock market crash was largely driven by high speed machines "trading" with other machines. Attempts to use super fast and efficient machines to replace slow and inefficient machines can have some scary implications. Michael Creighton's

—Kilobit (10^3 Bits/Second)
—Megabit (10^6 Bits/Second)
—Gigabit (10^9 Bits/Second)
—Terabit (10^{12} Bits/Second)

Figure 6.12.   Dimensions of Global Talk I

Medium Speed Data Channel	9.6 Kilobits/Second
Digital Voice	50 Kilobits/Second
Teleconference Channel	3 Megabits/Second
Digital TV	40 Megabits/Second
Advanced Communications Satellite	1 Gigabit/Second
TIUPILS	20 Gigabits/Second
HOLOVISION	1 Terabit/Second
GHIUDS	12 million Terabits/Second

Figure 6.13.   Dimensions of Global Talk II

book *Congo,* which cites a fictional General Franklin F. Martin's testimony to Congress, points a frightening picture of a future where computers make the life and death decisions in warfare because they are the only ones capable of "thinking" and "talking" fast enough.

As the hypothetical General Martin plausibly explains the historical evolution of computers and modern weaponry, there is little choice.

In 1956, in the waning years of the strategic bomber, military thinkers imagined an all-out nuclear exchange lasting 12 hours. By 1963, the ICBMs had shrunk the time course to 3 hours. By 1974, military theorists were predicting a war that lasted just 30 minutes, yet this "half-hour war" was vastly more complex than any earlier war in human history.

In the 1950s, if the Americans and the Russians launched all their bombers and rockets at the same moment, there would still be no more than 10,000 weapons in the air, attacking and counterattacking. Total weapons interaction events would peak at 15,000 in the second hour. This represented the impressive figure of 4 weapons interactions every second around the world.

But given diversified tactical warfare, the number of weapons and "systems elements" increased astronomically. Modern estimates imagined 400 million computers in the field, with total weapons interactions at more than 15 billion in the first half hour of war. This meant there would be 8 million weapons interactions every second, in a bewildering ultrafast conflict of aircraft, missiles, tanks, and ground troops.

Such a war was only manageable by machines; human response times were simply too slow. World War III would not be a push-button war because as General Martin said, "It takes too long for a man to push the button—at least 1.8 second, which is an eternity in modern warfare."

Since human beings responded too slowly," General Martin explained, "it was necessary for them to relinquish decision-making control of the war to the faster intelligence of computers. "In the coming war, we must abandon any hope of regulating the course of the conflict. If we decide to 'run' the war at human speed, we will almost surely lose. Our only hope is to put our trust in machines. This makes human judgment, human values, human thinking utterly superfluous. World War III will be war by proxy: a pure war of machines, over which we dare exert no influence for fear of so slowing the decision-making mechanism as to cause our defeat." (Creighton, 1980, pp. 280–281)

In addition to "super speed" problems of relating to computers, we will also have to cope with what has been called the *information overload problem*. Back in the 1930s, noted U.S. sociologist Robert Merton and Paul Lazarsfield experimented with rats. They overloaded the rats environment with too much noise and information. The rats became apathetic, they lost appetite, their sex drive went next, and then they died. Other than the sequence of events, the results are rather bleak. There is today a new word in the English language called *cacooning*. This new phenomenon of withdrawing from the hurly-burly atmosphere of information overload can be seen in home movie rentals, etc.

Professor Yasumasa Tanaka is one of the few researchers currently studying information overload. His research tells us some amazing things. He has found that information available to the public is expanding at a rate than is greater than 50% per annum, but human ability to assimilate more information is expanding at only about 3%. Tanaka's most alarming research concluded that, of all available options, the most successful information strategy to promote a new nuclear plant in a locality would be to "overload" all written, verbal, and broadcast channels to the extent that political apathy would accept almost any proposal. The broader implications of this finding when applied to other possible topics is to great concern.

Ironically, it may well be computers that can best help us deal with information overload. Personal computers that act as mailman, secretary, librarian, and file clerk will be necessary for coping with information diarrhea. The British Library's ASK program,[2] for instance, can help the researchers find exactly the information they want and filter out that which is superfluous.

The protection of personal privacy will be extremely difficult to achieve in such an environment. Electronic evesdropping is already easy. High resolution space photography is commonplace for surveillance. The so-called MediaSat project to provide "spies in the sky" for newsgathering purposes shows we can still go much further in this direction. Cross sorting and compiling of computer programs can tell an incredible amount about individuals. Many people feel that

---

[2] ASK stands for "Asynchronous State of Knowledge." This is a computer program designed to allow a researcher to "discuss" research needs with a computer, allow the computer to diagram the problem and ask clarifying questions, conduct a search of literature, and then help explain and interpret the results. See Belkin and Oddy (1981). For information about the "Last One," an English-instructable computer program that prepares basic utility-type computer programs through human commands, see "A Terminal Case" (1981).

recent moves to replace national census-taking activities with cross sorts of major government and commercial computer files is a dangerous move toward systematic computer monitoring of its citizenry. Already, law enforcement officers in New Mexico, Florida, and elsewhere have work release prisoners wear small transmitter bands attached to their legs the signals of which are received by an infrared device nearby. But new technology can go much further. A two-way high powered satellite could be used to monitor criminals or "subversives" by using a different frequency and two-way beeper like the one the new Geostar satellite system will use.

The problem of electronic crime is also a continuing and growing problem. A group of teenagers was arrested last year for giving erroneous commands and moving a satellite in orbit, accessing confidential medical records of severely ill patients, and altering charge card records. Professional electronic criminals have placed silent "computer bombs" in internal telephone data lines within banks' computerized accounting records. This "data bomb" contained a one-way filter that, if removed, served to "erase" days of transactions. The ransom to receive back the "erased" records could be substantial, for example, at 5% of the value of the total transactions. Today there are journals devoted just this issue. The problem is, of course, more than teenage hackers and professional electronic criminals. The ultimate threat could come from technoterrorists and even hostile nations. The full scale version of what could go wrong is outlined in Malcolm MacPhearson's book *The Lucifer Key*. (Incidentally, the INTELSAT satellites mentioned in that book are not intended to reference to real INTELSAT satellites which are different in purposes, design, and ownership from those in Mr. MacPhearson's book.)

What, then, is the pattern that we can see for the future electronic world, in terms of new developments, as well as significant problems to be solved? As Leonard Marks (1974, p. 173) former director of the U.S. Information Agency, put it:

> Global electronic networks . . . will pose realistic questions about information flow and cultural integrity. . . . These networks will move massive amounts of information through high-speed circuits across national boundaries. Moreover, they will be effectively beyond the reach of the traditional forms of censorship and control. The only way to "censor" an electronic network moving . . . 648 million bits per second is literally to pull the plug. The international extension of electronic mail transmission, data packet networks and information bank retrieval systems in future years will have considerably more effect on national cultures than any direct broadcast systems.

Key problems and issues related to our new technologies are listed in Figure 6.14. Alvin Toffler foresees a coming revolutionary transition that he calls the "Third Wave." This image is, I believe, a misleading one, because, unlike the waves from the ocean, each new Telepower wave of change is coming faster and

1. High Transmission Rates (up to 100 gigabits per second using laser and millimeter wave communications satellites, space platforms, advanced digital modulation & compression techniques).

   a. Information overload
   b. Obsolescense of equipment/accelerated
   c. Centralization vs. decentralization (both geographic & functional)

2. Rapid Growth of New Video Services & International Business Comm. Networks (videoconferencing, high resolution TV, multiple rastered TV, dedicated corporate networks for all forms of communication)

   a. Congestion of orbital arc in geosynchronous orbit & satellite proliferation
   b. Limited system interconnectivity & network breakdowns
   c. Concern over big vs. little network systems
   d. Developing countries' demands for New World Information Order. Reforms to status quo. Perceived extravagance in face of unmet needs of third world.
   e. Increased National trade protection in telecommunications area.
   f. National telecommunications monopolies' resistance to external "competition" and to customer-premise services.
   g. Emergence of electronic crime as a major problem (e.g., data banks, embezzlement, espionage).

3. Rapid Growth of Robotics and Artificial Intelligence in Global Industries Connected by Computer Networks. High-speed Machine-to-Machine Communications.

   a. Technological unemployment, underemployment, & skill loss.
   b. Redefining of economic values of work. New work ethics.
   c. National political resistance to automation within developing countries.
   d. New protection for personal privacy.
   e. Man-machine interface, psychological feelings of personal worth.
   f. Sabotage and terrorism.
   g. Questioning of Industrial Age Values.

4. Development of New Telecity and Extraterritorial Corporate States (e.g., free trade zones, such as in Sudan; floating, docked industrial units in offshore ocean locations; space platforms; moon colonies).

   a. Redefinition of roles and fiscal responsibilities of corporations and governments.
   b. Force of international & national laws for extraterritorial bodies, oceans, polar regions and subterranean areas of earth.
   c. Status of financing of new urban infrastructure, e.g., transportation, computer and communications networks; power and energy; utilities, etc.

Figure 6.14.   Developments by the Year 2000 and the Problems They Pose

5. Continued/Perhaps Increased Economic and Technological Gaps Between Developed and Developing Countries.

   a. War, terrorism, sabotage. Appropriate role of media and information flow.

   b. Appropriate international investment and marketing strategies.

   c. Need for new institutions and greater need for international law with effective sanctions.

6. Major challenges to Economic Growth and Prosperity Posed by the Limits of Earth Resources, Environment and Ecology. The "S-Curve" Limits of Growth.

   a. Effective strategies for space and the oceans.

   b. Relationships among countries, particularly developed, vis-a-vis, developing.

   c. Relationships among international agencies, governments and multinational enterprises.

   d. Need for effective incentives for "recycling economy" and conservation.

Figure 6.14   (cont.)

faster and the magnitude is ever greater. It would seem that we are facing a tidal wave of change. We are likewise living in an age of rapidly increasing "future compression."

Indeed if we were to review the 5-million-year history since Australopithecus man as being one single supermonth, the results would be startling (see Figure 6.15). Only at 9:30 p.m. on the last day of the month did man discover agriculture and invent the city. At 11:56 p.m. comes the Rennaisance and the creation of the scientific knowledge. This is the time of creation of what Pierre de Chardin characterizes as the "noosphere"—the age of scientific knowledge. The age of electronic computers, space travel, satellites, and television represents only the last 15 seconds of the supermonth. Within the next 60 seconds of supermonth time (or the interval from now until the 22nd century), the scope of change that mankind will experience will be stupendous, rewarding, frightening, and awe-inspiring. Perhaps we well see within this time Lunar and Mars colonization and terraforming of a planet or of a satellite of Jupiter. We should see hypersonic transport across the oceans either in vacuum-like tunnels with mag-lev technology aided by new "hot" superconductivity or by scramjet reusable rocket planes. But we will never see these or other marvels if we do not protect ourselves against political abuse of private information.

In some countries the terror of the World War II holocaust still lives on, particularly in terms of remembering how the wrong information in the right hands can do horrible damage. Today in Sweden national databases on Swedish citizenry are stored in computers, but computers where special precaution have been taken so that explosive devices can totally destroy the data in an instant if circumstances so warrant. There should be equal concern and precaution taken

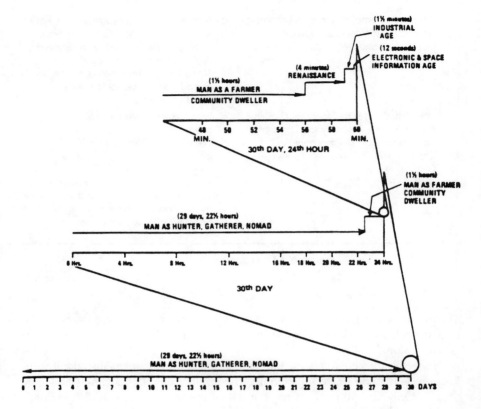

Figure 6.15. History of Man Depicted as a Supermonth

to ensure that no damaging use of information stored in government data bases is made.

CONCLUSION

In conclusion it would seem that the following value issues and associated possible corrective legislation with regard to those matters should be investigated as a matter of some urgency. Some attention has already been given to them in the past, but not enough.

1. Appropriate responses to issues related to technological unemployment; offshore electronic services; and international financial trading.
2. Assessment of the problem of information overload.

3. Potential abuses of electronic monitoring.
4. Assessment of the issue of cross sorting or cross compiling of computer files in lieu of national census.
5. Review of the problem of scale and speed in man-to-machine and machine-to-machine communications.
6. Unauthorized access to personal computer files and lack of access to ones own personal computer files.
7. Constructive use of personal computers and other electronic devices to maintain privacy; obtain wanted information and screen out unwanted information.
8. Trends and developments in electronic crime involving computers, artificial intelligence, and telecommunications systems.

REFERENCES

Belkin, N. N., & Oddy, R. N. (1981). *Design study for an anomalous state of knowledge (ASK) based information retrieval system*. London: The British Library.
Artificial islands could help Third World (1981, November 26). *New Scientist*, p. 603.
Creighton, C. M. (1980). *Congo*. New York: Alfred A. Knoff.
Marks, L. (1974). *International conflict and the free flow of information in control of the Direct Broadcast Satellite, Values in Conflict 66*. Palo Alto, CA: Aspen Institute Program on Communications and Society.
A terminal case for programmers. (1981, August 13). *New Scientist*, p. 410.

II
Corporate and Business Perspectives

chapter 7

Social Considerations in the Development of Telecommunications Policies

Joseph L. Daleiden

Director
Corporate Planning and Capital Deployment
Ameritech Inc.
Chicago, IL

Business men and women generally give little thought to philosophical issues. In a competitive, market-oriented society, monetary rewards are for action rather than contemplation. Indeed, the academic community is suspicious of the objectivity of a business person who dares to offer some opinions in an area customarily perceived to be the purview of scholars. Nevertheless, important philosophical assumptions are implicit in the positions of the various interest groups involved in matter of public policies, even if the antagonists may not always be conscious of them.

Consequently, in this chapter, I will discuss some of the philosophical implications of the basic controversies in the telecommunications industry today. I will begin by examining one view of the purpose of public policy in general, and then proceed to explore some of the issues with regard to regulation and competition as goals of public policy. Next, I will briefly review the history of telecommunications policy and focus on its more recent impacts on various segments of society. Finally, I will suggest an alternative direction for telecommunications policy which may be more effective in achieving the public policy goals outlined earlier.

THE PURPOSE OF PUBLIC POLICIES

Most people would agree that the purpose of public policy is to seek the public interest and to promote the common good. Yet, since the time of Plato, philosophers have found that attempting to define the term *good* is a formidable problem. Henry Sidgwick and G. E. Moore have summarized the difficulties in defining terms such as *good* and *right* (Sidgwick, 1907, p. 26; Moore, 1903, pp. 10–20). The problem arises when attempting to define good in some transcendental or absolute sense. The same problem underlies the "ought/is" controversy first examined by David Hume. From the perspective of an inanimate cosmos, it can never be said that something is good or "ought to be." Caring implies someone to care, and the someone can only be an inhabitant of our planet.

The term good is usually applied to something which will make people happy. Aristotle recognized that this was the customary usage and concluded that happiness is the proper end of humankind "because we always choose it for itself and never for any other reason" (Aristotle, 1953, Book One, VII). If happiness is the end of all human endeavors, values are instrumental rather than intrinsic, i.e., the value of anything is a function of its contribution to happiness.

If all people seek happiness, it is arguable that, for the most part, all people operate to maximize their own perceived self-interests. (I qualify this assertion, because sociobiologists such a O. W. Wilson may be right in their hypothesis that humans at times engage in certain genetically determined altruistic behavior which benefits their gene pool rather than the individual, Wilson, 1978, pp. 155–175.) Although Plato and Aristotle tried to escape the implication that happiness is a purely subjective experience, Epicurus recognized that that which makes a person happy is a matter of personal taste. But such preferences are both genetically and environmentally conditioned. One person derives happiness by teaching a course in economics, another by sacrificing his life in a holy war, and yet another by some form of strange ritual of sadomasochism. (Some might see an uncomfortable similarity in these three examples.) The point is that all three people are acting in their conditioned self-interest—all are seeking happiness.

Even though one might act from self-interest, and is, therefore, egotistically driven, I do not agree with Ayn Rand, who argues that humans are somehow psychologically and genetically free to objectively determine their own perceived self-interest (Rand, 1966). Nor do I believe that, in following perceived self-interest, one will necessarily attain happiness for oneself or anyone else. Plato thought that justice and happiness, which he equated to self-interest, could coincide. But Plato recognized that there is nothing intrinsic in the nature of man to assure this congruence. I would submit that creatures of unlimited wants who maximize self-interest may cause much unhappiness for all people, including themselves. Here I break with the classical economic tradition from Adam Smith to Milton Friedman which appears to rest on an assumption that ignores environmental constraints that set a finite limit on supply in both the short and long term.

An example of pursuit of perceived self-interest to the detriment of all is demonstrated by Garrett Hardin in the "Tragedy on the Commons" (Cox, 1985, pp. 49–64[1] In a somewhat fictitious example, Hardin explains that, in medieval England, where the pasture land was open to all, it was rational that each herdsman seek to continuously enlarge his herd. But since every herdsman thought the same way, and each operated solely in his own short-term self-interest, the common land was eventually overgrazed to the ruin of them all. In the 1920s, Aldo Leopold recognized the danger of the same situation developing in the Southwest of our own country (Leopold, 1979, pp. 131–141). Not only in issues of conservation, but in matters of public policies required for long term benefits (such as expenditures for alternative sources of energy, public education, or waste disposal), narrowly construed self-interest will result in policies detrimental to all in the long run. Bertrand Russell saw the necessity of balancing the self-interest of an individual with the needs of the rest of society. Russell wrote: "Our conduct, whatever our ethics may be, will only serve social purposes insofar as self-interest and the interests of society are in harmony. It is the business of wise institutions to create such a harmony as far as possible" (Russell, 1935, p. 24). As Leopold and those concerned with land-use ethics understood, it is also essential that the "needs" of society be brought into harmony with the physical carrying capacity of the environment.

Problematically, each person subscribes to a different set of values as a means of attaining happiness. One person might value security above all else as a prerequisite to happiness, while another enjoys adventure and unpredictability. One person wants to save the Redwoods so that he or she and future generations can take quiet, contemplative hikes, while another person would like to harvest them to be used as siding for his or her house. Resolution of those conflicts is the proper subject matter both of ethics and public policy.

Lester Thurow discussed the problem of balancing the desires of various interest groups in *The Zero-Sum Society* (Thurow, 1981). When one segment of society benefits at the expense of another, we are forced to contend with the difficult question of how to evaluate changes in net human happiness. The problem is how to avoid the trap of the classical utility theory of Bentham and Mill which logically leads to policies which maximize happiness for the majority at the cost of inflicting unhappiness on the minority.

John Rawls offers an interesting alternative to utility theory (or perhaps a refinement of it). Rawls suggests that, in deciding the rules of justice (social policies), we hide behind a "veil of ignorance" with regard to our specific circumstances (Rawls, 1971, pp. 136–142). In other words, we pretend that we do not know whether we are rich or poor, old or young, white or black, brilliant or learning impaired. We do not even known whether we will be part of this

---

[1] Actually, the English common land was managed so that such a disaster did not occur. Cox (1985) explains how the common land system disappeared for other, primarily economic, reasons. However, even if Hardin's account of the tragedy is apocryphal, it serves as a useful parable.

generation or a future generation. We then seek to determine the rules of justice which best contribute to the welfare of all members of society.

One problem with Rawls's approach is that, even if we could successfully suppress our biases regarding our personal demographic and economic circumstances, it might be impossible to avoid the bias inherent in our various psychological perspectives. The high-risk personality may opt for a winner-take-all world, while the less secure, more conservative personality would settle for the prospect of making much less in the way of rewards in return for an adequate safety net should a person fail. This caveat aside, Rawls rules of justice have much to recommend them.

Robert Nozick takes issue with the lexical structure introduced by Rawls whereby certain principles take absolute precedence over others; for example, justice takes precedence over efficiency and welfare. Nozick's attempts to develop a balancing structure whereby the right results of an act (or policy) are weighed against the wrong results (Nozick, 1981, p. 479). Interestingly, Rawls's general conception of his theory avoids the strict lexical ordering. Although Rawls feels this is a shortcoming, I think it is an improvement. Nozick's balancing structure then may be interpreted as a variation on Rawls' conception.

In deciding the best public policy, it is useful to ensure objectivity by hiding behind the veil of ignorance when assessing the consequences. But in making the selection among alternatives, we must also attempt to balance the impact of policies on the *relative* happiness of the various groups affected. Even if a policy increases the absolute welfare of all, if it improves the relative advantage of one group over another group, the policy may be perceived as unfair by the less advantaged group and, hence, may actually reduce the happiness of the group which receives the lesser benefit. This is so because people perceive their welfare in a relative, rather than absolute sense. For example, how many of us would be happy with a $5,000 raise (which might be considered delightful in itself) if we subsequently learned that all our colleagues got a $10,000 increase?

The difficulty comes in attempting to measure the trade-offs of alternative policies on the happiness of different social groups, and this is where Rawls's general formulation is more useful than Nozick's. Although Nozick discusses the use of a weighting scheme, it is not clear what his objective function is. He specifically rejects a utilitarian maximization principle and, although he gives a nod in the direction of some teleological values, he basically would like to adopt a deontological view consistent with his Platonic leanings. I would argue, however, that, in developing a set of normative rules, we would still need to decide on a fundamental objective (such as relative human happiness) as the basis for deciding which rules are best.

Therefore, if the purpose of public policy is to increase the happiness of all, and we perceive our welfare in a relative rather than absolute sense, social policies must improve the welfare of the worst off at least proportionally to that of the best off. Rawls' general conception of justice can be modified to support this position:

> All social primary goods (liberty and opportunity, income and wealth, and the basis of self-respect) are to be distributed equally unless an unequal distribution of any or all of these goods is to the [greater than proportional] advantage of the least favored. (Rawls, 1971, p. 303)

I inserted the qualifier *greater than proportional* to deal with the issue of relative happiness. In its modified form, Rawls's conception would permit those most favored in society to perform actions which would improve their advantage, thus providing incentives to maximize their performance, while insuring that, over time, such actions do not increase social inequalities.

Given this modified general conception of Rawls, specific policies must be formulated based upon empirical evidence of human nature, especially with regard to those psychological factors which create or satiate wants. Otto Neurath saw the need to ultimately treat the question of how to maximize happiness as any other scientific issue:

> One could . . . conceive of a discipline pursuing its investigations in a wholly behavioristic fashion as part of unified science. Such a discipline would seek to determine the reactions produced by the stimulus of a certain way of living, and whether such ways of living make men more or less happy. It is easy to imagine a thoroughly empirical "felicitology" (Felicitologie), on a behaviorist foundation, which could take the place of traditional ethics. (Neurath, 1973, p. 213)

One problem with an empirical approach to socioeconomic issues is that it requires tracking the reaction of various groups of people to a change in policy over long periods of time and trying to isolate the reactions which are specific to that policy from a myriad of other causal or coincidental relationships. It is a difficult and expensive task. But what other useful alternative is there?

In the past, both common law and religious doctrines have been used to establish normative ethical standards. Common law is grounded in a body of precedents that ultimately rests on philosophical or religious assumptions. But how can such assumptions by judged to be true unless there is some attempt to validate them through actual experience? Religious proclamations oftentimes are not even concerned with happiness in this world, or when they are, rely upon the pronouncements of leaders who claim to speak with God-given authority or upon speculations based on theological assumptions as to the nature and purpose of humankind. Moreover, religions usually put severe restrictions on those whose desires lie on the extremes of the bell-shaped curve of human behavior, even when that behavior may not conflict with other peoples' efforts to attain happiness.

COMPETITION AND HUMAN HAPPINESS

The difficulty, perhaps impossibility, of conditioning all persons to channel their desires in socially harmonious directions was recognized by Rollo Handy. Therefore, according to Handy, the task is not to make people "less interested than they are now in their own preservation, but rather to try to structure our social organizations so that one's own betterment is dependent on the betterment of others. In other words, we should try to organize human institutions so that, rather than one person's success being dependent upon the failure of others, everyone's success should be dependent upon the success of others" (Handy, 1973, p. 213).

Contrary to Handy's advice, American society conditions people to be more competitive. It is an unchallenged assumption that competition is the most effective way to achieve happiness. However, since unconstrained competition may fly directly in the face of the social harmony sought by Russell or Handy, the merits of competition should be clearly examined. Even if competition is the best means for maximizing production and improving the material well-being of a society, it does not necessarily follow that it is the most effective way of increasing human happiness. Moreover, even if some competition is desirable, there may be limits to the amount of competition that is desirable. The question has not been sufficiently studied.

One analysis by Nancy and Theodore Graves explored the long debated issue of nature vs. nurture with regard to prosocial development by comparing the behavior of adults and children of the Cook Islands and New Zealand. Not only did they study what prosocial changes occurred with maturation within the two groups, but, since so many Cook Islanders were migrating to New Zealand, they were able to see the effect of migration from one culture to the other.

Their testing procedure involved allowing each subject to choose one of nine pairs of cards which had a different number of pennies for both the chooser and an observing child of the same age and sex. The result showed that New Zealand children were far more likely to demonstrate *rivalrous* behavior (defined as either maximizing their own gain relative to the other or minimizing the other's gain in absolute terms), while the Cook Island children were much more likely to be *generous* (either maximizing the other's gain in absolute terms or maximizing it relative to one's own gain). Of particular note was the willingness of New Zealand children to sacrifice their own gain just so the other child would receive less!

Another surprising result was that the older Cook Island children began to display about as much rivalry as New Zealand children, even though the adult Cook Island population, especially in the more remote areas of the islands, did not demonstrate this rivalrous behavior. The cause could be traced to the teachers of the children. The teachers themselves tested as the most rivalrous occupation group on the Island, with two-thirds showing rivalrous, and no teachers exhibiting generous, behavior. Further investigation discovered that the "school setting . . . is structured according to New Zealand norms of individual work and competition toward personal goals, and discourages helping others to a common goal" (Graves & Graves, 1983, pp. 243–270). Other studies of rural Mexican children demonstrated similar results.

In his book *No Contest—The Case Against Competition,* Alfie Kohn adopts the radical position that competition is inherently destructive (Kohn, 1986). Specifically, he argues against four assumptions which he characterizes as unfounded myths: (a) competition is an unavoidable fact of life, (b) competition motivates us to perform our best, (c) competition provides the best way to have a good time, and (d) competition builds character. Although the Graves study and Kohn's arguments have merit, they are not compelling. I am particularly concerned that many of the societies which Kohn uses as examples of a noncompetitive ethic are, like Cook Island culture, facing assimilation, extinction, or exist at the caprice of more dominant, competitive societies. Unfortunately, it is indeed a Darwinian world. Nevertheless, it cannot be asserted á priori that competition is always good because it will somehow inevitably increase human happiness. Since, by its very nature, competition requires winners and losers, it will increase happiness for some people while decreasing it for others. Whether or not in the long run even the losers are happier than if they lived in a society characterized by an absence of competition is questionable.

In discussing the effects of competition, one must also consider the different consequences of intragroup and intergroup competition. When two basketball teams are competing, the players realize that, if their team is to win, cooperation within the team is essential to maximize the competitive position of the team as a whole. The same is true of firms. Kohn contends that competition kills productivity within a company. He explains that seeking to win is different than striving for excellence and advises managers to base rewards on the success of the group.

Not only do individual firms compete, but entire nations compete. The question then becomes, when is competition advantageous and when is it counterproductive? The focal point of competition is crucial to evaluating its results. If intraentity competition hampers the efforts of a team, firm, industry, or nation in its efforts to compete with an external competitor, the internally focused organization will be the loser. In Japan and China cooperation is the watchword; interpersonal competition is frowned upon. Even competing firms are much more willing to cooperate for the good of the nation than in this country. I would submit that it is at least possible that, in our efforts to avoid the negative aspects of the kinds of exploitative collusion which our antitrust laws were designed to prevent, we have permitted the pendulum to swing too far the other way. Under the assumption that competition will in every instance lead to enhanced consumer welfare, and failing to recognize the international nature of competition today, the courts and regulators are virtually promoting competition as an end in itself, irrespective of potential social consequences.

The discussion thus far can be summarized as follows:

While it is impossible to determine what constitutes the good in an absolute sense, we can subjectively decide to use the term good to describe any action which increases human happiness. Each person seeks to increase his or her own happiness by pursuing action in his or her own self-interest. In so doing, conflicts arise between persons. Social policies are developed in an effort to reconcile those conflicts. The ideal is a harmonious reconciliation, but that is not always possible. In developing these policies a guiding principle should be that the benefits of the policies (their contribution to happiness) should be distributed equally or, if unequal, to the greater than proportional advantage of the least favored. It is generally an unquestioned, yet problematic, assumption that the promotion of competition somehow always benefits society. However, not only must competition result in losers as well as winners but, unless competition is correctly focussed, it can result in a dissipation of energy and resources in internal scrimmages which should be directed more profitably against an external challenger.

Given this cursory philosophical perspective, I will next explore the consequences of the introduction of competition into telecommunications and the subsequent divestiture of AT&T. How did these events affect the various interest groups involved? Who were the winners and losers? Can we envision a telecommunications policy which incorporates the Rawlsian concept of justice?

CAUSES OF COMPETITION AND DIVESTITURE

Before speculating on the future of telecommunications, and suggesting policies which best meet the needs of the various constituencies affected, it is necessary to provide a brief historical perspective as to the confluence of forces which brought us to the present situation. That the exact make-up of the telecommunications industry today was partially a political happenstance is suggested in Steve Coll's account of the breakup of AT&T, *Deal of the Century* (Coll, 1986). Politics aside, however, the industry was in for a major restructuring due to technological trends and changes in the global economy, which will be reviewed in a moment.

First, let me provide a thumbnail sketch of the origins of the what was the world's largest regulated monopoly, the Bell System. Perhaps the reader has seen those 1910 photographs of telephone poles in urban areas which are supporting literally hundreds of telephone lines. The lines were strung by several competing telephone companies, each serving its own subscribers. Each company also had its own switching facilities, so calls could not be switched between companies. If a caller wanted to ring someone subscribing to another telephone company, he or she would have another telephone on his or her desk. Clearly this was inefficient and expensive. Moreover, from a societal perspective, the duplicative facilities were a waste of capital resources. It was apparent that having only one company to switch local calls was socially advantageous, and, therefore, the practice of granting exclusive operating franchises was adopted.

A similar logic was employed in the Federal government's decision to allow AT&T to retain its monopoly of intercity facilities. The Federal Communications Commission (FCC) was established to ensure that AT&T did not take undue advantage of its monopolistic position and to foster the goal of a low-cost national telecommunications service.

Often, the stated rationale for pricing was the classic notion of a utility curve. Since the phone of the businessman was thought to be more valuable to his or her concerns than that of the resident customer, it seemed fair to charge him more. And, since the alternative to long distance service was the slow, non-simultaneous service of mail or telegraph, telephone service was obviously considerably more valuable to long distance users. Public policy was predicated on the ability to pay principle rather than the market solution, which undoubtedly would have precluded many people from having telephone service. In short, the regulators were interested in maximizing the happiness for the greatest number, reflecting the utilitarianism of Jeremy Bentham and John Stewart Mill rather than the social Darwinism implicit in Adam Smith. As a consequence, an elaborate system of subscriber cross-subsidies evolved: from urban to rural, from business to residence, from long distance to local service, and from vertical services (such as premium telephone sets, touch tone, extensions, etc.) to basic service. In setting prices of telephone service for specific subscriber groups, the FCC and local regulators were primarily concerned that phone service was affordable to as many people as possible; the long run public goal was universal service.

Up until the mid-1960s, rapid productivity improvements in the Bell System permitted the Bell companies to constantly lower their rates. (My studies of Illinois Bell's total factor productivity indicate that the company was improving at the amazing rate of 4.8% per year, compared to the U.S. average, at the time, of about 2.0%). This meant that, in general, telcos had very happy consumers; and happy consumers meant happy regulators. However, the stimulus to inflation sparked by the Vietnam War, and the Johnson Administration's effort to finance it by printing money, soon changed the regulatory climate. The Bell companies found that their productivity growth could not keep pace with inflation and the higher returns demanded by their investors; hence, they were forced to seek rate increases. In some localities they were turned down cold or given only a fraction of what they requested.

The regulators also were suspicious that the telcos might be deliberately inflating their rate bases by adding unnecessary telephone plant. Since regulation sets prices to permit an adequate return on capital investment, there indeed appeared to be an incentive for the Bell companies to boost aggregate profits by increasing plant and equipment expenditures irrespective of a specific project's payback. Recognizing this possibility, but not having the means to directly evaluate the economically appropriate amount of capital investment, the regulators in some states were extremely reluctant to increase prices.

In an effort to maintain adequate returns in the face of falling margins, some of the Bell companies resorted to overly stringent capital programs. Their decision was ill-timed. The U.S. economy was entering the information age, when the geometric growth in voice and data transmission would put a huge strain on the network. In certain key areas, such as the financial area of New York City, the strain caused a near collapse of basic service. The Bell System was accused of neglect and indifference stemming from its monopolistic position.

At the same time, it became apparent that the convergence of information processing and information transmission technology was producing a fundamental change in the very definition of the telecommunications industry. It was obvious that either information processing companies would integrate into the telephone industry, or telephone companies would attempt to integrate into the information industry. The new central office switching machines were simply large computers which could achieve significant economies of scope if they were also used to process data. This prospect, which appeared so exciting to the telephone industry, gave the computer manufacturers nightmares. Moreover, the intransigent stance adopted by John DeButts over the issue of connecting customer-provided terminal equipment to the Bell network did little to assuage the fears of the information industry that the telephone companies would leverage their control of the network to prevent the computer manufacturers from entering telecommunications.

The regulators had mixed feelings about all this. At the local level, some regulators saw that revenues from new services could be used to hold down rates on basic telephone service. Other regulators thought that the telcos might use the basic service revenues to subsidize their incursion into the information industry, thus forcing the ratepayer to bear the cost and risks of the new ventures with little hope of reaping the benefits.

At the federal level, the FCC shared these concerns. The Department of Justice (DOJ) was also convinced that AT&T would try to push its monopolistic nose into the competitive arena of information services. The raison d'être of the commercial sector of the DOJ was to foster competition. Ever since Theodore N. Vail and President Woodrow Wilson worked out the Kingsbury Commitment of 1919, which gave legal status to AT&T as a regulated monopoly, the Justice Department had opposed the agreement as antithetical to the free enterprise system. It believed that competition would always provide greater economic benefits than a regulated monopoly could. Even during the last antitrust case, the DOJ never attempted to prove that AT&T had illegally set prices below cost, only that regulation could not prevent AT&T from setting prices to destroy competition (Coll, 1986, p. 192).

The final player to enter this complex drama, and in the opinion of many the precipitating factor, was MCI. From the time when it was a small provider of microwave services between Chicago and St. Louis, to when MCI became a major national network, its CEO demonstrated a political acumen which AT&T, with all its size, or perhaps precisely because of its size, severely lacked. In establishing his business, Bill McGowan recognized that he could make the system of cross subsidies, developed to provide universal service, work in his favor. Whether or not he planned from the beginning to develop a second long distance network is a matter of conjecture, but that he was able to take advantage of the uneconomic pricing structure thrust on the Bell System is indisputable.

Not only could MCI underprice AT&T, since it had no local exchange or unprofitable long distance routes which it was forced to subsidize, but, equally advantageous, MCI did not have to engineer for peak demand times. It could engineer transmission facilities to maximize utilization of its network, and then dump excess call volumes back on the AT&T network, thus exacerbating AT&T's peak engineering problems as well. When AT&T sought to prevent MCI's practice of serving only the most profitable routes, it quickly found itself in court. The evidence to support my contention that the system of cross-subsidization was essential to the success of MCI is shown by the financial history of MCI since divestiture. Since then, AT&T has been permitted to lower interstate prices closer to cost (a 36% decline between 1984 and 1988), and the burden of supporting the local network has gradually been shifted to the end users through increased access charges. The result was an erosion of MCI profit margins to such an extent that it had to rely upon IBM to bail it out until MCI could complete its own long distance network. Even then MCI could not have gained enough market share to survive had it not been successful in getting the courts to force the local Bell companies to modify every one of their switches—at the rate payers' expense—to provide the same 10-digit dialing for long distance service which AT&T could offer.

The confluence of these several forces ultimately led, in circuitous fashion, to the breakup of the Bell System. However, the eventual divestiture agreement was not the result of a logical policy based upon the principles such as utilitarianism or Rawlsian justice. Rather, the basis for telecommunications policy was aptly summed up by Steve Coll:

Precious little in the history (of U.S. v. AT&T) was the product of a single, coherent philosophy, or a genuine reasoned consensus, or a farsighted public policy strategy. Rather, the crucial decisions made during the 1970's and early 1980's were driven by opportunism, short-term politics, ego, desperation, miscalculation, happenstance, greed, conflicting ideologies and personalities, and finally, when Charlie Brown thought that there was nothing left, a perceived necessity. (Coll, 1986 369)

CONSEQUENCES OF THE INTRODUCTION OF COMPETITION AND DIVESTITURE

Did the dismantling of the Bell System and the introduction of competition lead to greater happiness for all segments of society? Not surprisingly, there were winners and losers.

Bell System Employees

For many employees of the Bell System it opened up new opportunities and new challenges. The winds of change blew open the doors of the highly structured, predictable, even stodgy Bell System and reinvigorated many a manager in mid-career. Some of those who left the Bell Companies were surprised to learn that they had very marketable technical or sales skills. As AT&T and the new Regional Bell Operating Companies (RBOCs) sought to reduce their force, managers were oftentimes able to tuck a sizable pension or early retirement incentive into their pockets and then walk across the street to a competitor and begin a new career.

However, for many other managers and craft persons, the consequences of competition and the breakup of the Bell System were cataclysmic. These were the employees who joined the Bell System with the belief that their mission was to provide the highest quality of telephone service, and that if they performed their job satisfactorily, they would be virtually guaranteed lifetime employment. They were now thrust into a new world where all bets were off. They felt betrayed. Regardless of how well they did their job, their employment was at the mercy of the marketplace and the decisions of top management. For those who transferred to AT&T at divestiture the situation was particularly distressing. As international competition cut away at AT&T's markets, particularly in manufacturing, layoffs of both craft level and management employees accelerated. In every area of its manufacturing business, AT&T has been faced with dwindling market shares. It has been particularly hard hit in sales of PBXs and central office switches. The AT&T share of the PBX market plummeted to 26% by 1988 (Gartner, 1989), while in only 3 years its market share of new switches dropped from 80% to 49%. As a consequence of all the turmoil, medical departments at both AT&T and the RBOCs have witnessed a sharp increase in illnesses related to job stress.

On the other side of the ledger, there have been hundreds of thousands of new jobs created in the myriad of new companies formed to get a bite of the telecommunications pie. Unfortunately, many of the promising new careers turn out to be illusionary, since the market shake-out has already begun. Also, even before divestiture, the FCC's policy to permit any registered terminal equipment to be hooked onto the network resulted in soaring growth of telecommunication imports. This situation accelerated after divestiture as the Bell Companies sought additional suppliers of central office equipment. Since AT&T was the sole U.S. supplier, and since the RBOCs were precluded from manufacturing their own equipment, they were forced to import switches. The result was a further deterioration of the U.S. trade balance in telecommunications, with the deficit reaching $3.5 billion by 1988.

Investors

Shareowners who held AT&T stock at divestiture generally did well if they converted their stock into a proportional number of AT&T and BOC shares. As the table below shows, returns to RBOC shareowners between January 1984 and December 1989 more than doubled the S&P 500 average, while AT&T share-owners slightly exceeded it:

S&P 500	168%
AT&T	195%
Avg. of 7 RBOCs	351%

An analysis of the impact of competition and divestiture on the investors of the myriad of other companies in the telecommunications industry is beyond the scope of this chapter. Suffice it to say that the fortunes of the winners and losers were multiplied by several orders of magnitude relative to the Bell System shareowners.

COMPETITORS

Divestiture has had little effect on the interconnect market, which had already become extremely competitive since the time of the FCC's 1968 Carterphone Decision. The financial fate of the U.S. competitors entering the telecommunications marketplace has been subject to wild swings from success to failure. Many of the manufacturers of terminal equipment, especially at the low end of the market which was already over-crowded before divestiture, have conceded the field to Japan, Korea, and Hong Kong. Even in the area of PBX manufacturing, profit margins have been whittled away. Some of the fastest-growing interconnect companies have discovered that their only salvation has been to be acquired by someone with deep pockets and a desire to integrate information processing with switching and transport. The sale of Rolm to IBM and later to Siemens is a case in point. The interconnect industry has learned that, in the absence of artificial regulation on pricing of central office services and court-imposed limitations on product offerings, the economies of scale and scope would make RBOC central office services competitively advantageous for the majority of the business market.

Many of the executives of the old Bell System had no objection to the idea of interconnection if they were allowed to compete fairly. The real issue for them was whether interexchange competition was really viable in the absence of any cross-subsidies which created pricing anomalies to be exploited. In other words, the question was, and still is: "Is the network a natural monopoly?" I submit that the evidence, both before and since divestiture, suggests it is impossible to decree competition in a market where the economies of scale would preclude it. Despite the Regional Bell Companies spending billions of dollars to remodel their networks to provide equal access to the Interexchange Carriers (ICs), only one major interexchange company, AT&T, has been consistently profitable.

There are many who believe that, unless AT&T deliberately goes easy on its competitors in an effort to gain and maintain the benefits of deregulation, the industry will revert to its natural monopolistic status. However, given the protection that the other ICs received while they constructed and recovered the costs of their network facilities, this may not be the case. While the ultimate success or failure of competition in the interexchange market is still open to question, the evidence to date suggests that, at best, the IC market will be an oligopoly with AT&T acting as the price leader.

Consumers

What has all this meant for the poor baffled consumers? Are they happier with competition after divestiture? The results of one study showed that 80% of the people surveyed were satisfied with their telephone service before divestiture. Since the breakup:

64% thought it was a bad idea,
11% were not sure, and
25% approve (Coll, 1986, p. 367)

The problem with the studies on consumer satisfaction is that they do not segment the market. The impact of introducing competition to the telecommunications industry and breaking up the Bell System can only be assessed by examining the consequences for each consumer group. A heavy user of long distance services would be happy to see the 36% drop in long distance rates. But many of those who used the telephone primarily to make local calls received a rude shock. The impact on their telephone bills depends on where the subscribers live. If they are fortunate enough to live in an urban area of the Midwest, they may have experienced a rate reduction. But if they had the misfortune of living in a high cost rural area of the South or West, they might be cursing the wonderful folks who decided it was in the nation's best interest to end the system of cross subsidies. In many areas the price of local telephone service has more than doubled since divestiture, and this doesn't include the skyrocketing cost of installation and repair. Moreover, these increases would be higher still were there not still substantial local subsidies. But it is the awareness of the requirement that local telcos must "overcharge" their large users to subsidize the occasional user that is encouraging large users to install and manage their own networks, a function which telcos could do at much less cost due to their economies of scale.

In general, the business customers, especially large businesses, are happy with the greater number of options they receive under competition, including the potential to merge information management and transport in their own network. On the other hand, the residence customers on fixed incomes have just one more worry with which to contend.

THE HIDDEN COSTS OF COMPETITION

There are a few additional costs to the consumer which I would like to touch upon before discussing the future of telecommunications policy in the U.S. Economists implicitly assume that two of the assumed major results of competition—greater innovation and diversity—are unalloyed benefits. But innovation and diversity involve some very subtle costs and, consequently, may not always increase happiness.

When purchasing services, such as provided by a restaurant or the theatre, I have my relatively brief moment of enjoyment, pay for it, and that's that. On the other hand, when purchasing a capital good such as a camera or a VCR, although I pay for it up front, I expect the benefit to last over a certain period of time. The business person recognizes this time dimension through the amortization process, whereby a certain amount is charged to depreciation expense each year over the economic life of the equipment. Economic life is the length of time it takes before the net discounted cash flows generated by a new piece of equipment will make it economically advantageous to replace the old equipment. With rapid technological innovation, equipment lives are becoming shorter and shorter. A shorter life means higher annual amortization charges.

Ultimately, all innovation costs in telecommunications must be borne by the marketplace. However, how to allocate those costs becomes an issue, because in our industry much of the equipment is common to all users. The telephone cable going down an alley may serve both residence and business customers, high volume and low volume users, some wanting Plain Old Telephone Service

(POTS) and others wanting the most exotic data transmission services available. Whether costs are allocated based on usage, incremental demand, or some other criterion becomes a crucial point. In the old days the decision was relatively simple. With POTS service, usage was roughly proportional to cost and so provided a fair basis for allocation. Prior to the introduction of competition the telcos could, to some extent, control the introduction of new technology. Therefore, they could make certain that all costs of existing equipment were recovered before introducing the new technology.

In competitive markets, the introduction and life cycle of new technology is ultimately controlled by the marketplace. If a business fails to recapture its equipment investment by the time an innovation occurs, the business probably never will, since competition will force it to adopt the innovation, cut prices, or loose sales. Businesses set their depreciation schedules accordingly. However, in the regulated world of telecommunications, depreciation schedules are set by regulators whose primary interest is to keep prices to the residence customers as low as possible. Therefore, regulators naturally wish to keep depreciation rates low. But, if a telco fails to recapture its costs of an investment prior to introduction of a competitor's innovation, the investor loses the earnings on that investment. Here is a classic trade-off of happiness between two interest groups—the consumer vs. the investor.

Diversity also has a hidden cost. The more products we introduce to our homes, the more time it takes to research and compare the features and prices of each competitive alternative, and the cliché "time is money" is certainly true. Are we reaching the point where the bewildering variety and number of options have become a source of frustration rather than happiness? I think many elderly people seek to simplify their lives with fewer choices.

How does this relate to telephone service. One of the most frequent complaints since divestiture was not that service deteriorated, or that prices increased; rather, it was that consumers were forced to make choices which many of them did not want to make. Many did not want to buy a phone and worry about obsolescence, or choose between price and quality from dozens of alternative products. Nor did they want to select an interexchange carrier (IC). When customers were forced to choose their IC, many stuck with AT&T simply because they chose not to choose.

I am not an intellectual Luddite throwing my wooden shoes into the wheels of progress. I simply wish to suggest that there are costs associated with innovation and diversity which are too often ignored by economists. People may be happy with what they have, and giving them something new or different may not increase happiness, or, more likely, it may increase it for some while decreasing happiness for others.

Finally, little attention has been paid to the issue of the social costs associated with the huge amount of additional investment required by introducing competition to the telephone industry. To enable the other ICs to be competitive with AT&T, the Modified Final Judgement against the Bell System required that it modify or replace its central offices so that all IC customers could dial long distance with the same number of digits as AT&T customers. The cost of providing this "equal access" to all long distance callers has been tremendous, perhaps as high as $17 billion by the time the switching offices are converted nationwide. Moreover, there is a possibility that, in the near future, the telcos will have to provide local access for intralata calls as well. An even more complex issue is the requirement that the telcos unbundle their network services into separate offerings such as calling number identification, billing, central office switching features, facility usage information, etc., and offer each element for sale separately. Such unbundling requires new accounting systems, new network management and control systems, and may, paradoxically, result in more, rather than less, regulatory oversight.

A second set of investment costs arise from the enormous amount of excess network capacity being constructed. If true competition is to exist for network services, it requires that each of the IC competitors provide a virtually ubiquitous network. But there is already a ubiquitous network in place. What social purpose do a second and third network serve? If the answer is that they will foster competition, it begs the question, why did we need competition in the first place? If the answer to that question is that competition was required to drive down prices, it could be argued that regulation could have accomplished the same end at far less cost. As I have said earlier, the reason that long distance rates were high was to provide a subsidy for local service. If the regulators wanted to reduce the price of long distance service, all they had to do was end the cross subsidies. As for those who contend that the relatively small staffs of regulators are incapable of controlling the gigantic telecommunications industry, the answer may have been to simply increase the FCC staff. When one considers the enormous number of personnel and capital employed by the competing ICs (MCI alone expends 3 billion dollars in expense and 1 billion dollars of capital annually), compared to the FCC's paltry $90 million budget, which includes the regulation of all TV, radio, etc., it is at least arguable that regulation could be more socially efficient than competition. Again this is an empirical question.

In evaluating the cost and benefits of competition, we must consider the national opportunity cost of the investment dollars being used by the ICs to build their duplicative networks. Like everything else, the total amount of U.S. investment capital is limited. To the extent that it goes to financing telecommunications equipment, other investment opportunities are foregone and the cost of all investments is driven up. Even in the near future, the excessive demand for telecommunications investment might retard investment in other areas— particularly social investments. In a nation which is woefully underinvesting in its public infrastructure, we can ill afford to waste precious capital resources.

Finally, there is the argument that competition fosters innovation. I have always felt this was a curious argument to bring against an organization which had 19,000 patents and was responsible for inventing and/or developing the vacuum tube, transistor, silicon chip, laser, cable television, and even radio telephony—the technology utilized so effectively by MCI. What made the argument so specious was that the Bell System could have been encouraged to bring new technology to the marketplace faster by simply permitting the Bell companies to shorten their depreciation schedules to reflect economic depreciation rather than physical depreciation—an argument that the telcos still have trouble selling to the regulators. In point of fact, the cornucopia of new products and services which was supposed to be the result of imposed competition has yet to occur.

Aside from some innovative pricing schemes, I fail to see the promised plethora of innovative products from the new ICs. Surprisingly, it is in countries such as France, where the state-run telephone company faces absolutely no direct competition, that the development of information services such as videotex has taken off. In this country, on the other hand, introduction of cellular phone service was delayed for 10 years so that competition could develop. My concern again is that federal regulatory policies may be more driven by ideological considerations than by the desire to satisfy the consumer.

FUTURE DIRECTIONS IN TELECOMMUNICATION POLICY

Today it cannot be denied that there is competition in the telecommunications industry, and that it is growing. The Bell System has been broken up, and Humpty Dumpty cannot be put together again. There are, however, two issues which still must be faced:

1. Should the telcos remain regulated, and if so, by what mechanism?
2. To what extent should the Regional Bell Holding Companies be permitted to diversify?

In this short treatise I cannot do justice to either issue, but allow me to offer some considerations. With respect to the first issue, the regulators are in a box. Unless they want to regulate all the competitors—and that would be admitting that pure competition is not in the consumer's best interest—the need for the RBOCs to respond to the incursions of competition will force a dramatic change in the nature of regulation. The only way the regulators can avoid the ICs extending their networks all the way to the large end users is for the RBOCs to be permitted to compete on price and service. This means prices to end users must reflect actual incremental costs. The result will be reduced prices to high volume, low cost users—as warranted by the economies of scale—and higher prices to the low volume, high cost customers. If the regulators don't permit this, the large users will bypass the local telco by hooking directly with the ICs, or else build their own network. In either case, the regulators must end up raising rates to those remaining on the local network.

Recall that I proposed earlier that the purpose of any public policy should be to increase happiness. However, the Rawlsian concept of justice contends that it isn't enough that a policy increases happiness for some at the expense of others who are disadvantaged to begin with. Therefore, I would submit that proper telecommunications policy should strive to protect the individuals most likely to suffer the negative consequences of competition—the low volume users, especially those with low incomes.

With this objective in mind, what are the regulators' options? They could assess all providers of telecommunications services a charge to provide a subsidy to those who might otherwise not be able to afford basic service. Conversely, the government could completely end regulation and use the tax system or transfer payments as the means of subsidization. (To some extent this is occurring today, since small rural phone companies can borrow capital dollars at subsidized rates from the Rural Electrification Fund.) But either of these two solutions would be very complex to implement and would require an additional government bureaucracy to administer.

A more practical solution is for regulators to adopt a social contract whereby the telcos guarantee basic service at a minimal fee with future price increases at some rate below the rate of inflation. This means at least some amount of cross-subsidization would continue to exist. The quid pro quo for these price ceilings would be for the regulators to allow flexible pricing for services where competitive alternatives do exist, and to drop their efforts to control the RBOCs rates of return. Eliminating rate of return regulation would provide the necessary incentive for the RBOCs to improve operating efficiency. What possible incentives are there today to become more efficient if all the benefits of productivity improvement flows to the customers and none to the owners of the business?

There is one other crucial concession that regulators and the courts (aye, there's the rub) must make if the RBOCs are going to be able to continue to subsidize service to their residence customers. The RBOCs must be allowed to provide those subsidies from profits gained from the economies of scale and scope generated from the sale of enhanced services. It is absolutely nonsensical to bar the RBOCs from offering the entire panoply of information services.

Three arguments have been advanced by Judge Green and others as reasons why the RBOCs should be prevented from entering new markets such as information services. First, it is maintained that the RBOCs will seek to use their profits form local exchange services to subsidize entry into new markets. By means of these subsidies they could then underprice and ultimately drive out competition. Second, since the RBOCs are regulated on basis of rate of return, it is feared that they will seek to increase aggregate earnings by expanding the rate base through additional services, even when the new services are unprofitable. Third, it is alleged that the RBOCs exercise a bottleneck on local exchange service; therefore it is assumed that they will throw up artificial restrictions or demand excessive prices to choke their competition.

The first argument assumes that the RBOCs are willing to incur losses through predatory pricing and also run the risk of prosecution under the Robinson-Patman Act, which forbids pricing under cost to drive out competition. Why the RBOCs would undertake such a substantial risk for uncertain long term gains defies the imagination.

The second argument, that the RBOCs would seek to uneconomically expand their rate base through the addition of unprofitable services, is easy to remedy and should have been cared for years ago. The answer is to regulate prices rather than rates of return. By regulating prices the BOCs would be discouraged from undertaking any venture which is uneconomic, since to do so would reduce profits. Another, more cumbersome approach, would be for the regulators to insure that all new services cover long run incremental costs.

The third argument—concern that the BOCs maintain a network bottleneck—is being met by the FCC requirement that they open their network to ensure equal accessibility to all users. However, even without FCC action, this argument fails. As Howard P. Marvel explains, "the source of value of information service is the information itself, and the BOCs are not particularly well positioned to be information gathers" (Marvel, 1987, p. 63). Perhaps more significantly, the primary business of the BOCs is the sale of network usage. To the extent that information services stimulate greater usage, it is in the RBOCs' best interest to welcome as many competitors into the market place as possible.

In the final analysis, the information industry is not any different then any other emerging market. To protect certain competitors by forbidding others market entry is a form of protection which will deny the marketplace the benefits of the most cost effective players. The RBOCs may possess a competitive advantage because their central offices are simply very large computers and, since the local network is ubiquitous, the incremental cost of offering a variety of information services may be quite low—so low that, perhaps, many other businesses wishing to offer these services could not compete. In a free market that might happen. Those with an ideological commitment to creating and sustaining competition do not appreciate that competition and free markets are not synonymous. In a free market, if there are substantial economies of scope to be achieved by a company leveraging one service off another, it is an economic crime to prevent this social benefit.

What Judge Greene seeks to forbid by his administration of the Consent Decree, the regulators want to preclude by spurious accounting techniques. By forcing the RBOCs to price to cover fully allocated embedded costs, rather than incremental costs, they are making the same error which caused the railroads to lose the battle to the trucking industry despite the true economic advantage of rail for intercity transport. The same ideological mindset which forestalled the introduction of cellular phones in this country for 10 years after they were successfully trailed in Chicago is now preventing the telcos from offering the benefits derived by integrating information transport with information processing. It is ironic that countries such as France, whose telecommunications systems lagged woefully beyond the U.S. for so many years, now have taken the lead in providing information services to their residential customers.

In the field of manufacturing, it makes no economic sense to preclude the BOCs from competing. Every foreign PBX manufacturer is a competitor to the Bell companies' CENTREX service. In most cases, the international competitor has a protected home market. With its embedded cost covered by its home market, it can seek to gain economies of scale by invading the U.S. market.

I've already noted the devastating impact of the growth of international competition on the balance of trade in telecommunications. AT&T argues that, if the BOCs were freed from the manufacturing restrictions, they might choose to form joint partnerships with foreign competitors. The resulting increase in imports might result in a still larger balance of trade deficit. Admittedly, this is possible, but it is beside the point. In essence, AT&T's argument is a plea for protection against foreign competition. Economists have long ago demonstrated that, with the possible exception of developing countries, any form of protectionism works to the disadvantage of all members of the world economy. By restricting imports, protectionism would force Americans to pay a higher price for telecommunication service, thus reducing their ability to spend that differential on other goods and services. The result is a lower standard of living. A far better policy for the U.S. is to pressure our trading partners to eliminate their trade barriers and open all markets to worldwide competition.

Finally, although it is doubtful that the telcos would want to broadly compete in an already overcrowded interstate market, there may be some niche markets which they could enter. If they cannot, it should be the marketplace, not a bureaucracy, which makes that decision. At any rate, there is no chance that they will pose a serious threat to AT&T's dominant position.

SUMMARY AND CONCLUSIONS

What does this discussion of telecommunications policy have to do with happiness? In all the issues I have touched upon, telecommunications policy may work to the benefit of some and the detriment of others. In each instance, the goal of the policy makers must be to evaluate the trade-offs on the relative happiness of affected segments of society.

Rawls' theory of justice holds that the principle of justice is lexically prior to the principle of efficiency. It might be argued that telephone service is today a basic necessity which our society wishes to provide to all its members. Therefore, the first aim of telecommunications policy is to promote the universality of basic telephone service at affordable rates. This can be achieved through price caps. However, to avoid discriminating against investors in telco stocks, regulators must permit a return on investment commensurate with the additional risk inherent in the introduction of competition. This can best be achieved by abandoning rate of return regulation in favor of price caps. In so doing, regulators provide the profit incentive for the telcos to continually improve their efficiency.

Competition must always be considered as a means to the end of increasing human happiness, rather than as an end in itself. Free market policies may or may not lead to more competition. The goal of telecommunications policy should not be the promotion of competition per se, but efficient markets. Whether the result is ultimately a competitive market, an oligopoly or a monopoly should be decided by the marketplace. If a monopoly is the natural consequence, then regulation is an appropriate response.

In general, human welfare is enhanced by the introduction of new services. However, simple multiplicity of identical services may involved significant and, possibly more than commensurate, social costs in terms of dublicative investments, and consumer confusion. If, however, an open market is believed to be the most effective way to stimulate the offering of new services, no potential competitors should arbitrarily be prohibited from the marketplace. The RBOCs should be encouraged to enter any market even if, perhaps I should say especially if, their economies of scale or scope would give them a cost advantage over competitors.

REFERENCES

Aristotle. (1955). *The ethics of Aristotle.* (J. A. K. Thompson, Trans.). New York: Penguin Books, Ltd. (Original Publication 1953, Allen & Unwin).

Coll, S. (1986). *The deal of the century.* Boston, MA: Houghton Mifflin.

Cox, S. J. B. (1985, Spring). No tragedy on the commons. *Environmental Ethics, 2.*

Graves, N. B., & Graves, T. D. (1983). The cultural context of prosocial development: An ecological model. In J. D. L. Bridgeman (Ed.), *The nature of prosocial development.* New York: Academic Press.

Handy, R. (1977). Ethical theory, human needs, and individual responsibility. In P. Kurtz (Ed.), *Moral problems in contemporary society.* Buffalo, NY: Prometheus Books.

Kohn, A. (1986). *The case against competition.* Boston, MA: Houghton Mifflin.

Leopold, A. (1979, Summer). Some fundamentals of conservation in the southwest. *Environmental ethics.* Albuquerque, NM: John Muir Institute for Environmental Studies.

Marvel, H. P. (1987, November). *Organization and competition in telecommunications: An idiosyncratic view.* (The National Regulatory Research Institute, Occasional Paper 12.) Columbus, OH: Ohio State University.

Moore, G. E. (1903). *Principia ethica.* Cambridge, England: Cambridge University Press.

Neurath, O. (1959). Sociology & physicalism. (M. Magnus and R. Raica, Trans.). In A.

J. Ayer, (Ed.), *Logical positivism*. New York: The Free Press, A Division of Macmillan Publishing Co., Inc.

Nozick, R. (1981). *Philosophical explorations*. Cambridge, MA: The Belknap Press of the Harvard University Press.

Rand, A. (1966). *Introduction to objectivist epistemology*. New York: New American Library. (Original publication in the "Objectivist", July 1966–February 1967).

Rawls, J. (1971). *A theory of justice*. Cambridge, MA: The Belknap Press of Harvard University Press.

Russell, B. (1961). *Religion and science*. Oxford, England: Oxford University Press. (Original publication in the Home University Library, 1935.)

Sidgwick, H. (1981). *The methods of ethics*. Indianapolis, IN: Hackett Publishing Co. (Original publication by MacMillan and Company Limited, 1907.)

Thurow, L. C. (1981). *The zero-sum society*. New York: Penguin Books. (Original publication by Basic Books, Inc., 1980.)

Wilson, E. O. (1979). *On human nature*. New York: Bantam Books. (Original publication by Harvard University Press, 1978.)

Wilson, E. O. (1989). PBX markets: 1988 review, 1989 prospects. *Strategies in telecommunication services* (Newsletter, M-590-878.1). Stamford, CT: Gartner Group, Inc.

chapter 8
Transborder Data Flow: Global Issues of Concern, Values, and Options

Riad Ajami
Professor and Director of International Business
International Business Program
College of Business
The Ohio State University

Transborder data flow is quickly becoming the most important issue facing international trade in services. The major forces driving it to the forefront include:

- The shift to a world economy based on service rather than industrial production,
- The internationalization of markets, and
- The growth in the use of computer processing and telecommunications.

For the world economy to continue to grow, and for the world's resources to be used most efficiently, it is critical that the transfer of information continue across national borders unhampered by restrictive regulations. As nations realize the power that lies in the processing and transmission of information, however, the advantages of improved world efficiency are often counterbalanced by other national and indigenous values such as autonomy, diversity, and more parochial considerations.

After reviewing the types of transborder data flow, the economic and non-economic issues of concern will be examined. The chapter will then outline the broader impact of transborder data flow restrictions on international trade.

Two principle forces anchor this study, a political imperative and an economic one. Autonomy, pluralism, and diversity (ideals indigenous to the political imperative) are the guiding values for national entities. Effectiveness, efficiency, accuracy, and timeliness (economic imperative issues) are the underpinning of strategic alternatives for the multinational corporate entities (MNEs). In this light, the study will assess the conflicting views on transborder data flows from the perspective of both national governmental entities and

126

multinational corporate actors, and look into the development of competitive strategies by each in dealing with the rapidly changing transborder data flow environment juxtaposed to determine the strategic alternatives and preferences of the actors.

TYPES OF TRANSNATIONAL DATA FLOW

Much data flow daily through international networks, not all of which are subject to the restrictions we wish to impose on transborder data flow. Therefore, it is important to understand the different types of data and data flows.

Data is categorized by *content* and by *flow pattern*. We have four major types of data flow content (Novotny, 1980).

1. *Operational Data* are international data flows which support organizational decisions and sustains certain administrative functions. MNEs, for example, use such information to coordinate geographically dispersed business functions.
2. *Financial Transaction Data* represent the information resulting in credits, debits, and transfer of monies. These are distinct from operational data containing financial information. While the unrestricted flow of financial data permits convenient financial agreement, it also makes it difficult for governments to control currency speculation.
3. *Personally Identifiable Data* contain information relating to credit and/or medical histories, criminal records, employment and travel reservation, or simply names and identification numbers. Personally identifiable data may also appear in operational and financial transaction data.
4. *Scientific and Technical Data* include items such as experimental results, surveys, environmental or meteorological measurements, and economic statistics. Bibliographic databases and software which process raw data are also made available to the international scientific community through computer communication systems.

Participants

As well as understanding the types of data involved, it is important to know the nature of the entities involved in transborder data flow. Virtually any and every entity that transfers information across borders is a participant in transborder data flow activity. The small manufacturer which sends a brochure to a foreign inquiry is as much a participant as the giant MNE and the nation-state. Participants in transborder data flow can be categorized generally as:

- Sovereign States
- Other governmental entities
- Private and quasigovernmental organizations
  - private and common communication carriers
  - data processing service bureaus
  - MNEs
  - transnational associations

These different participants have diverging views of how to use, promote, exploit, regulate, and monitor transborder data flows. These differences give rise to several issues of concern.

Issues of Concern

The issues surrounding transborder data flow are economic, technological, cultural, sovereign, and multinational. By its very nature, transborder data flow is an economic issue requiring investment in technology and personnel to be operational, creating employment and facilitating international commercial transactions as outputs. The relevant technologies include computers, satellites, telephones, microwaves, and a host of others, all of which affect and are affected by economic considerations.

Transborder data flow is a cultural concern as well. The transmission of personally identifiable data affects individuals' rights to privacy (OECD, 1980). Intercultural differences with respect to personal privacy are thus affected. The cultural concerns of latecomers, as well as those of other Western industrial democracies in Europe, are also manifest as the international media, television, and multinational corporate entities move us closer towards a global village.

On the other hand, transborder data flow is a sovereignty issue. The nation-state, by the nature and purpose of its mandate, desires to control information crossing its border and to minimize the political and economic costs of such interaction (Mowlana, 1986). The nation-state is empowered to promote the well being of its constituents. This may be in conflict with the economic well being of the multinational corporate player. Technological changes and advances have contributed, and will certainly continue to contribute, to the controversies and will make monitoring transborder data flows more complicated (UNCTC, 1981; Sardinas & Sawyer, 1983).

Finally, transborder data flow is a multinational corporate concern. National regulations on transborder data flow restrict operational flow patterns and impede organizational decisions and administrative functions in multinational network flow and the ability of MNEs to use information to coordinate geographically dispersed entities. Moreover, restrictions on transborder data flow are akin to nontariff trade barriers. These barriers limit the competitive advantage of individual firms in an increasingly competitive global market.

IMPACTS OF TRANSBORDER DATA FLOW ON INTERNATIONAL TRADE

To gain knowledge about several transborder data flow issues, the direction and type of data flow can be analyzed. Figure 8.1 presents a simple overview of transborder data flow patterns (Turn, 1979).

From the sketch above mapping out the patterns and potential flow patterns of transborder data flows, we can look at and analyze the potential impacts of transborder data flow on international trade.

The Emerging Global Economy

International trade is manifested through global economic growth. There has been a dramatic shift in the world economy. While exact percentages vary, it is estimated that service industries in the United States account for almost two-thirds of the total employment. Revenues from global trade service activities are

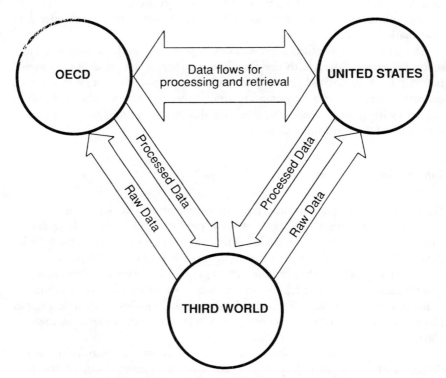

Figure 8.1. Directions of Transborder Data Flow

estimated to be almost over $400 billion, approximately 20% of total trade.

The single most important development in the expansion of trade in services has been the merging of computer and communication technologies. This union has opened the door for tremendous efficiencies in processing and production. It has also added a new dimension to the world economy by creating world trade in telecommunications, data processing, and other information services. It has also enabled developing countries to attract additional industries, particularly from MNEs, because of lower communication costs.

This fundamental change in the world economy has also been mirrored by the change in the cast of major players in transborder data flow. Historically, there were only five companies involved in transborder data flow between the U.S. and other countries. These international record carriers (IRCs) were Western Union International, RCA Global Communications, TRT Telecommunications Corp., ITT World Communication, and FTC Communications. This group has been dramatically expanded to include, not only IRCs, but also state-run post, telegraph, and telephone monopolies (PTTs), and private domestic communications carriers. The explosion of the use of computers and telecommunications equipment has been largely responsible for this decentralization. Deregulation of the telecommunications industry in the United States has also created substantial expansion in the number of players in the transborder data flow market.

The expansion of information services in particular and the entire services sector in general, along with the changing transborder data flow environment, has not gone unnoticed. Now, information technology is considered by all countries to be a strategic commodity vital to their economic growth. As often happens with strategic resources, there are also increasing cells for protection of domestic information industries.

Information as a Commodity

We live in an information-intensive society, where information sometimes becomes more important than both physical products and services. Information can be bought, and is, in fact, traded. This raises the issue of treating transborder data flow as a commodity which should be taxed as it crosses borders. A number of countries already practice discriminatory pricing, and have imposed tariffs and questionable government regulations in order to protect national information industries from the effects of transborder data flow. Transborder data flow out of Denmark is subject to export licensing, and in France, transborder data flows are classified according to retail value.

Moreover, as Figure 8.1 shows, there is an asymmetric relationship in the direction of transborder data flows. Raw data are exported out of the host country, processed, and imported back into the country. This leads to a loss of potential jobs in host countries—both in the LDCs as well as in many of the Western-world developed countries.

Separation of Production and Decision Making

The geographic separation of the decision-making process and the production which occurs in some MNEs is feared by many countries and intertwined with the issues of transborder data flow. No country wants to become too dependent on political and economical groups outside its borders. Even European countries feel they are too dependent on U.S. computer and communications technology. This might change as large projects, such as the European space program (ESPRIT), attract U.S. MNEs to locate research facilities in Europe.

Transborder Data Flow Regulation

A major area of conflict is in the regulation of transborder data flows. The United States has taken the position that there should be an unfettered "free" flow of information regardless of national frontiers or borders. Other OECD countries have a much different outlook. These countries look to safeguard the privacy of the individual through government agencies or monopolies. These agencies have been empowered to regulate the flow of data which is proving increasingly important to the worldwide growth of the service industries.

The viewpoint taken by the less developed countries is that development and use of transborder data flows should be restricted to protect each nation's international sovereignty and economic options. Determined not to allow this vital aspect of today's multinational corporations to go unrestricted, the Third World sees transborder data flow regulation as a possible means to more equitable share, on a global basis, the increasing economic gains which are resulting from the telecommunications revolution.

These three world views tend to clash as reflected in the following areas:

1. The commercial and economic impact of transborder data flows and the movement to regulate such flows through both tariff and nontariff barriers as well as direct taxation.
2. The concerns of individuals in the use of personal data.
3. The ongoing political battle being fought in various organizations worldwide and the public image and legal implications of these battles.

When considering if a type of data flow complies with local and international regulations, it is necessary to consider the directions of transmission and the location of users and storage equipment (databases). For the MNE, it is important to remember that a type of flow and content legal in one country might be illegal in another.

Political Barriers to the Free Flow of Data Across Borders

As mentioned earlier, in this chapter we are analyzing the outlooks of the two major driving forces in the transborder data flow arena, the economic forces and the political forces. From the economic perspective, an arena where companies are allowed to compete freely against each other is ideal. No company would have any particular advantage, and every company would have equal access to the same information. At first glance, this would seem to be ideal from the political aspect as well. Companies create jobs, and commerce brings in revenues. In the inherently international area of data transfer, free competition would bring competitive equality for national business.

Intuitively, from the economic perspective, barriers to transborder data flow do not make sense. Political entities, i.e., governments, however, are legitimized by providing services other than simply bringing in and allocating revenues for a country. A primary purpose is protection of individual rights and privacy from the forces of foreign institutions which seek to impose constraints deemed not to be in the best interests of the population which placed the extant government in power. Even the centrally controlled governments protect their citizens from improper manipulation by foreign institutions. Nations are thereby moving quickly to erect just the sort of protection on informational advantages or resources that have historically been associated with more durable goods. So the stakes of this battle over transborder data flow are quite high.

All nations today realize the critical importance of transborder data flow in their prospects for economic growth. (Less developed countries (LDCs) are particularly sensitive to this reality.) Consequently, the increasing restrictions on transborder data flow are a result of increased national sensitivity to the strategic importance of transborder data flow to national economic growth, a desire to develop on indigenous information infrastructure, and a fear of continued exploitation are thus introduced which typically resemble classic trade barriers and include the following:

- Discriminatory pricing of data transmission services.
- Mandated use of national public data networks.
- Denial of leased lines or restrictions on use.
- Local content laws requiring processing on data in country of origin.
- Restrictions on the importation of equipment, spare parts, and software.
- Emerging policies which may lead to customs duties and value-added taxes on information as it enters or leaves a country.

The commercial communications links are affected by two main types of transborder data flow regulations. The first restricts the free flow of information by limiting the categories under which transborder data flow is acceptable. This deals for the most part with privacy concerns, especially as laid out by the

OECD. The form of transborder data flow regulation is the more direct form of barrier in that these types of regulations which restrict data flow are directed at the MNEs for the expressed purpose of regulating international commerce.

One of the largest contributors to impediment to free flow in the OECD has been the host country PTTs (post, telephone, and telegraph), which control the communications sector in most European countries. Accorded favorable treatment and encouragement by the national government, these public monopolies erect nontariff barriers which hinder the free flow of data, eliminating or substantially reducing the competitive advantages of the multinational corporations.

Specific Examples of Value Issues in Conflict

Within the past several years, countries have begun implementing their own specific restrictions to transborder data flow. While not all of these restrictions have been documented, there are several more publicized examples that have had considerable effects upon the restricted and restricting countries.

Some specific examples include the 1980 Canadian Banking Act, which mandates that initial processing on all banking activity being done in Canada before transmission to foreign countries. In addition, restrictions are also placed on the processing of copies outside of Canada. This law was enacted for bank supervision. However, many feel that the real motivation behind the statute was the amount of Canadian data being transmitted to the United States. A Business International study, for example, found that the vast majority of Canadian transborder data flow involves flows from subsidiaries in Canada to the parent, primarily in the U.S. (Robinson, 1980).

The Canadian Banking Act of 1985 required banks that do business in Canada must carry out their data processing of customer records in Canada. If this restriction is violated, the Canadian Minister of Finance can suspend processing outside the country if he believes that the processing is not in the country's best interest.

Many countries are considering usage-sensitive rates for leased lines to encourage use of internal processing facilities, and West Germany began charging MNEs high volume-based public and private line tariffs. Continued moves in this direction could increase processing costs substantially.

Taxation is becoming an issue as countries such as France express interest in taxing data transported across their borders ("Madec Expects," 1982). While this is potentially a source of revenues to nations, the taxing of transborder data flows is designed to encourage the use of domestic facilities than to raise money and indicates the importance of this service to nations.

Finally, some European countries have restricted drug companies from transmitting specific information on individuals who have participated in tests on new pharmaceutical products.

It is important to note that, while some of these restrictions are similar in form, there is no unity on specific laws. Each country, in an attempt to deal with value conflict (preserving autonomy and national sovereignty) vis-á-vis global unification strategies of multinational corporate entities, has devised its own system for dealing with such problems.

The newly industrializing countries and other LCDs manifest other "unique" concerns about TDF. Governments are concerned about the vulnerability of vital national data to access by foreign governments or corporations. Some governments perceive dangers to their cultural integrity due to their inability to control the flow of information to their own citizens. For example, satellite broadcasts and access to satellite services are volatile issues in many LDCs. The unlimited reception of satellite broadcasts is seen as a threat to national sovereignty. It is perceived as a vehicle for the dissemination of propaganda, commercial domination, and cultural intrusion. The number of communications satellites which can be in orbit at the same time is finite. Nations with insufficient economic or technological resources to put their own satellite in space at this time fear they will be shut out by space limitations in the future. This places them at a continued economic disadvantage. They find themselves in a dependency relationship, lacking the tools or power to compete. In a report to the International Communication Problems Study Commission, specialist Delbert Smith notes that, given the spacing needs in geostationary orbit, and the reality that there is only room for 180 satellites like the INTELSAT V, it is, therefore, perceived as indispensible to developing countries that they reserve orbital parking places, even if they are not ready to use them (Masmoudi, 1984). So, even if these countries developed their communication network to the point where they could create an integrated network for information flows, the critical question of available orbits may still remain unanswered.

Typically, the U.S. is thought of as the most free environment for data processing, a philosophy strongly supported by the current administration. Some have even suggested that the U.S. could become a "data haven." However, philosophy and action often differ. For example, the Reagan Administration ordered Dresser Industries to stop its French subsidiary from producing parts for a Soviet pipeline. Consequently, the company changed the access code to its main computer in Pittsburgh, preventing the French from obtaining needed design data. Dresser-France subsequently lost business with Australian and Japanese firms because of its inability to obtain needed information from the parent's U.S. database.

Even the Olympics provide an example of U.S. interference with transborder data flow. Cuba requested some satellite hours to transmit Olympic coverage. However, the State Department invalidated ABC's deal with Cuba and imposed technical restrictions on the network so it could not transmit via satellite.

In less developed countries (LDCs), the pressure to adopt protectionist trade policies concerning services and telecommunications goods is even greater than in the OECD.

For example, Brazil has a wide array of transborder data flow regulations and prohibits the transmission of data for outside processing unless that processing capability is not yet available in Brazil. It also requires that MNEs use locally manufactured equipment. Chase Manhattan's application to purchase a computer from a U.S. vender was rejected, since the computer was not produced locally (Miller, 1986).

In addition, the adoption of teleinformation plans in the hope of protecting telecommunications industries may in actuality undermine the long-term economic interests of LDCs. By burdening the multinational corporation with dramatically increased costs, the end result for LDCs counterparts could be less productivity. The advocates of a freer international information order suggest that all concerns (advanced industrial economies and those of the newly industrializing countries) would be better asked where ease of access to information prevails. The question then becomes, should additional concessions be allowed in the international transfer of data? Again there is the question of which values should prevail. Complicating this question is the stage of development of the industries of the NICs and the LDCs. Being in earlier stages, these industries are not as efficient as their counterparts in the industrialized world, therefore they are not as competitive. Should they be protected and nurtured at the cost to their national constituency, or should the larger, more efficient corporations be allowed to provide for the needs at a lower cost? Is the long-run benefit worth the short-term cost?

Obviously, there are some significant world economic implications from restrictions on transborder data flow. These barriers, like all trade barriers, disrupt economic activity, hurt the growth of the world economy, stifle improvements in domestic economies, and negatively affect growing competition that has developed in some countries. From the MNEs viewpoint, the continued uncertainty for the future of transborder data flow regulations makes it extremely difficult to plan. Also, to the extent that competitive positions of firms are increasingly dependent on transborder data flow, continued chaos in the variety of regulations will make it difficult for MNEs to prosper and grow. It should not be problematic to ascertain that the conflicts between these two forces are difficult, at best, to resolve. To begin examining potential resolutions to these conflicts, an examination of the "values" which are affected by and affect the decisions on transborder data flows is in order.

Values: Global Diversity Vis-á-vis Centralization

Global diversity and pluralism are cherished values to many, and scale and giantism are often seen as detrimental to these potentialities. Multinational corporate giants (in the telecommunications fields and other sectors) attempt to reduce environmental complexity by wielding power and dominance. Economic concentrations, and the extraterritoriality of global communication giants, are troubling to geographical fixed national elites in host societies as well as to social system designers interested in providing a richer and more intricate pattern for human welfare and experience.

THE GLOBAL CORPORATE SYSTEM AND THE REMOTENESS OF POWER

Multinational managers appear to rule over vast enterprises with extensive power and resources. To national elites they appear unaccountable, irresponsive, and subject to very few mechanisms of control, internal or external, while affecting significant control over the lives of many.

Accountability and responsiveness ideally include participation by those affected by the decisions effectuated. Participation is never an absolute and people's lives will always be touched by decisions made by others. What most people have in mind is a continuum. Institutional systems are more or less participatory. Some systems tend to lean toward the participatory end of the continuum, while others are more centralized and exclusionary. The corporation tends toward a degree of centralization of power unacceptable to its critics, rendering quite difficult, if not nearly impossible, a meaningful degree of participation in its decisions. As such. It is a natural target for those who advocate that people and societies have the right to participate in the decisions that touch their lives. This is a simple yet sacred democratic maxim. Those who violate or deny it are faced with the burden of justifying their policies. To national elites, the corporate technostructure is seemingly remote and incapable of portraying even a facade of democracy and accountability.

Other problems visited upon the multinational corporate system arise from the decentralized nature of the international system. The world's central political and economic institutions are notoriously weak and ineffective. Though most of the world's problems are global in scope, authority remains concentrated at the level of the nation-state.

To the critics of the existing world political system, George Modelski's (1972) conceptualization of the global political order as a "layer cake" model is suggestive of the demise of the nation-state as a viable sovereign entity. In this context, the nation-state layer is overgrown at the expense of the global and local layers. Rhetoric about its obsolescence not withstanding, the nation-state is neither dead nor ailing. It is alive and doing quite well, if well, that is, designates its strength rather than its compatibility with human welfare.

The multinational corporation, domiciled in a home country, straddles the world's boundaries. The question of accountability to various constituents therefore becomes particularly touchy. The corporate argument, that it seeks to be a good citizen in all the host countries in which it operates, does not, by and large, suffice in the international context. In a strictly national economy, the political order may fail to regulate or check the power of domestic corporate interests, but the doctrine of national sovereignty can always be invoked, and there is, potentially at least, the promise of asserting the primacy of collective national needs and interests over those of the corporate system. The fit between the political order and the economy, and the coextensiveness of their boundaries which prevails in a domestic political order domiciled within its boundaries, does not, however, obtain in a system involving large multinational corporations.

Debates concerning this phenomenon have emphasized the challenges and dilemmas that the mobility of the multinational corporation poses for host countries. A number of important policy issues are raised in this context for providers and users of telecommunications technologies: barriers to entry, data hording, access to data, satellite use, and rights of early arrivals in space. Important as these issues tend to be, the overriding issue to the host society is the loss of national control or, more precisely, the erosion of its power to control and thus a compromise of its values. Governmental elites may dwell upon the economic costs and disadvantages or benefits that accrue from the presence and operations of multinational subsidiaries, but it is the issue of national control that lies at the heart of the concern.

As for the national societies as a whole, the opposition to the multinational corporate system is motivated by the desire on the part of individuals and communities to control their own destinies and to be masters in their own house. It is here that the centralizing designs of the multinational corporate system come in conflict with the desire for autonomy on the part of individuals and communities. No economic calculus can capture, or come to terms with, what might be called a spirit of localism, national pride, or patriotism.

Individuals want to reduce their vulnerability to forces that lie outside their control. Security and well-being tend to be associated with control or at least the potential for control. And I must emphasize "at least the *potential* for control." Iranian peasants may have had very little control over what the Ayatollah Khomini did in Teheran, but the boundaries of the policy were important enough, and political mythology, as well as a bit of political truth, held the promise of influencing the Ayatollah. Folk political assumptions, as a last resort, portrayed him as a figure with trusteeship over the welfare of the society. This ideal extends to all national leaders, elected, appointed, or ascended by divine right. No similar legacy can be claimed by foreign multinational corporate executives. Historical tradition gives the national leader a degree of charisma and legitimacy which could never be matched by a multinational technocratic system based in a far-away place. When all else fails, Iranian peasants and their urban countrymen can march on Teheran and demonstrate. It is viewed as impossible for them to make it to London or Delaware for similar expressions of their discontent.

Corporate executives, and the analysts who see matters from their perspective, may dwell upon the globalism of the multinational corporate system, but to some there is an element of feudalism to the entire system. Centralized multinational corporations are perceived as the modern equivalents of absentee-land-lords. They are remote, inaccessible, and appear as outsiders. The nature of legitimacy in the modern world decrees that the source of authority has to be coexistent with its impact.

Because of the primacy of national identity and the importance of interaction between multinational corporations and nation-states, the desire for autonomy is portrayed as expressing itself in strictly nationalistic terms. Samuel Hunting-ton, though, puts forth a more persuasive argument. The issue is not as much one of nationalism as it is of hostility to remote centralized power and the way it undermines the foundations of local communities (Huntington, 1973).

Advocates of the multinational corporate system dwell upon the inadequacies of the nation-state model. But their design, the giant centralized multinational enterprise, hardly improves upon the political model. The corporate model lacks the degree of accessibility (admittedly marginal and in desperate need of sub-stantial increase) that national political systems afford. George Ball (1968) rightfully may see the world's archaic political institutions failing to come to terms with contemporary needs and aspirations and may welcome the advent of the multinational corporation, but more than an indictment of the nation-state system is needed if one wants to prove the merit of the corporate model. Charles Kindliberger (1970) may argue with enthusiasm that the "nation-state is just about through as an economic unit," but those concerned with global welfare would still argue whether the shift in power away from it and toward the corporate system would be a particularly worthwhile development.

Here, I am reminded by how a folk Arab saying presents an undesirable choice: it is a choice between going deaf or going blind. Thus stated, the choice is not a particularly good one, and I guess it ultimately depends on whether one prefers not to hear or not to see.

Obviously a more adequate choice than the one between deafness and blind-ness is in order. The same goes for the choice between a world dominated by states or one dominated by corporate giants. Other alternatives are clearly in order. When all is said for and against both systems, the fact remains that citizenship in a political order still confers more rights and power than mem-bership in a corporate system. The governing elites of the state are still a good deal more responsive and accountable in the long run to the demands of their stockholders than is the multinational corporate elite. To not be unduly cynical about the entire matter, national elites are more territorially bound than their corporate counterparts. Though the former can leave the country, and often do when crises erupt, they have less flexibility on the whole than the latter, and their commitment to the local constituency is, by the nature of their legitimacy, inevitably greater.

When mobile corporations operate in a system of sovereign states, the prob-lem of jurisdictional conflict and jurisdictional avoidance results. This problem is in critical need of resolution, particularly if world benefit is a goal of the interaction between the nation-state and the multinational corporation. This therefore calls for a degree of centralization at the global level of the interests of the nation-states—a certain concentration of power *upward* to global political institutions with international regulations. Another problem however, that of the need for local autonomy, calls for a degree of decentralization. The demand for autonomy calls for a distribution of power *downward* to small nations, small economic units and local communities. In our conceptualization, both trends are compatible with one another.

A Model with Compatability

The model we have in mind for satisfying both needs is an economically federal model of sorts at the global level. It would satisfy the need for *central guidance* and *local autonomy* at the same time. More fundamentally, an adequate central guidance system would provide for a system of global chartering of corporations which operate multinationally. This system would contribute to easing and ameliorating the tension, disruptions, and perceived injustices associated with the existing asymmetric system wherein corporate headquarters are concentrated in a few strong, central societies. Under this federal umbrella corporate entities would be chartered and domociled in a center society. Local autonomy and diversity can be encouraged and strengthened. There is no inherent tension between the imperatives of central guidance and local autonomy. Indeed, in the case of weak communities and actors in the world system, a central guidance system is essential for their welfare and participation in issues that impinge upon the lives of the members.

The free-for-all laissez-faire assumptions that the international economy operates under are neither realistic nor would they be desirable if they were to obtain in the world system. They are not realistic inasmuch as they fail to portray the way an economy dominated by corporate giants tends to operate. Furthermore, like all laissez-faire assumptions in situations where such assumptions do not really obtain, they are, at best, ways of rationalizing and legitimizing why some gain and others lose in social transactions. A system of central guidance would strengthen the hand of small nations as well as subnational communities.

A system of central guidance would, and should be expected to, confront and deal with the way that large corporations are internally governed. We could foresee the possibility, under a central guidance system, of corporate boards of directors becoming general assemblies which reflect the various constituencies and nationalities that come within the reach of the corporate system. A more public and universal recruitment of directors could conceivably get around the way the corporate system excludes from its decision making those whose lives it profoundly affects. Corporate policies under such a system could be infinitely more legitimate and less resented than they presently are. With the benefit of experience with outside (nonmanagement) directors in the domestic context, we should be able to avoid the prospects of such individuals playing a merely symbolic role. They should be informed and backed by experts familiar with the intricate details of corporate operations. This system of governance is particularly urgent, and for that matter quite feasible, in the case of large and important global industries such as telecommunications.

The impetus on behalf of a system that combines central guidance and local autonomy, by its very nature, should come from all component parts of the world economy. Center nations are no longer as confident as they once were that they can control the activities of large multinationals and skew the distribution of costs and benefits in their favor. The interest of periphery societies in a system of central guidance is rooted in the difficulty that they face when they attempt to check and regulate multinational corporate power.

Though the impetus for such a system can come from various entry points and regions in the world economy, there are also differing steps and strategies that are more suitable for particular countries by virtue of their position in the international system than they are for others.

Some Strategies for Periphery Societies

In a rather gross and encapsulated form, the dilemma of periphery societies in the contemporary international economy can be attributed to the simple and undeniable fact that these societies have to operate in a world that is, for the most part, made by others. Since all webs of relationships and arrangements are structured in a manner that favors those who spin them, the relative weakness of newcomers is built into the structure and dynamics of global relationships. As latecomers to a modernized world, periphery societies enter an arena where others have decided how games ought to be played and which skills ought to be rewarded. Unlike the early arrivals, their politicoeconomic choices are vulnerable to the whims, designs, and policies of the powerful actors and trends in the international system. No country in the world, be it a firstcomer or a latecomer, has developed in total independence and isolation from external trends and influences. What differs about the experience of most latecomers is the extent and degree of their vulnerability to the muscle, lure, and machinations of others.

To recognize and proceed from the vulnerability of the periphery societies to dominant trends and forces in the international system is not, however, to argue that these societies are helpless pawns to be moved at the discretion of others. Liberation and autonomy begin, for individuals and societies, when the external constraints on one's action and freedom are recognized, accurately read, and finally confronted.

Standard writings on the relations between periphery societies and the structures of international finance and production are generally fond of comparing the GNPs of such societies and the sales of the largest corporations. The statistics tell us something, but they also mislead us. Latecomers are not without choices, options, and leverage in the world economy. Creative ways of organizing their domestic economy and conducting their international transactions can minimize their vulnerability to the more powerful and enhance their capacity to have a reasonable degree of control over the forces that affect their welfare.

The roots of the phenomenon of "dependency" are mostly intellectual and mental. We can dwell on its politicoeconomic and military dimensions, but it is ultimately a mental framework: some people innovate, and others emulate them; some are trailblazers, and others are followers. As long as individuals and nations go down trails and routes pioneered by others, it is doubtful whether they can cultivate their own creativity or have a reasonable degree of autonomy.

The basic and most fundamental step in periphery societies is primarily intellectual: Where do they want to go and how do they get there? If the framework within which these questions are raised and answered emanates from within these societies, then the prospects for a reasonable degree of autonomy would improve. If, on the other hand, other people's models and games are emulated and played, these societies are bound to be at a major disadvantage in dealing with others. Contrary to folk wisdom, it is absurd for a turtle to race a rabbit; the turtle should, if it wants to confront the rabbit, choose a game that capitalizes on its hard shell. For latecomers, the principal challenge is in re-visioning their future and priorities, and defining them in ways that capitalize on their own capabilities and posit goals that they can realize.

Resolving the Conflicts

The complexity of the problems arising from the TDF will continue to increase as the use of international communications networks increase. The sooner we move towards an international agreement, the better. Attempts have already been made. In 1980 OECD issued "Guidelines Governing the Protection of Privacy and Transborder Data Flows of Personal Data," which was adopted by 18 of its 24 member governments. Also, the Council of Europe issued "Convention for the Protection of Individuals with Regard to Automatic Processing of Personal Data." The latter is enforced whereas the first is meant as guidelines. Both regulations cover the privacy aspect of TDF. Additional international regulations are needed.

One problem is agreeing on regulations for TDF; another is how to control adherence to the regulations. How can data flow over international communication links be checked for illegal contents? Other unanswered questions are emerging, such as: how can data be quantified as tangible products for tax purposes, when neither sender or receiver charge a price for the data?

Moreover, MNCs must deal directly with the problems posed by TDF regulation. It is important to establish a dialogue with those states and organizations establishing the regulations. MNC's must make TDF a priority because of importance it has to their operations. All relevant information on host country TDF considerations and policies as related to corporate policy must be scrutinized. Multinational corporations must reassess their own current TDF policies and positions internationally in terms of future TDF actions within the host country. It may be necessary in some situations to make TDF a primary responsibility of a high-level management person if the company faces a restrictive and/or variable regulatory environment. Every attempt must be made to understand the motives behind the regulation. This would involve a study of cultural, economic, historical, political, and social factors of host societies.

If both the capabilities of late comers to use data and their access to the international market are improved sufficiently, the social and economic conditions of all will be improved. This reduction in disparity between the prosperous countries and the "new arrivals" to the information revolution is an outcome all should encourage.

Conclusions

In this chapter the economic promise of the emerging communications technologies through the free flow of data across national boundaries was explored. Benefits such as efficiency, accuracy, and timeliness in the transfer of data can help commerce flow more smoothly, creating additional benefits for the consumer as well as the provider of economic activity. These benefits are not without constraints and costs. Limited (scarce) resources such as the "parking spaces" in the geocentric orbit create the necessity of trading off free access on a competitive basis for "equal" access to all participants. The options available are many, each with different allocations of the benefits and costs.

Among the costs are the changes that inevitably occur in the value structure of the interacting societies. We have explored these value structures as well. With insight and forward thinking, we can create a harmonious balance between the two opposing imperatives and extract the most benefit globally at the least cost economically and socially.

REFERENCES

Ball, G. (1968, October). Making world corporations into world citizens. *War/Peace Report*, p. 207.

Huntington, S. P. (1973, April). Transnational organizations in world politics. *World Politics*, pp. 341–342.

Kindliberger, C. P. (1970). *Power and money.* New York: Basic Books.

Madec expects TDF dividends for France. (1982). *Transnational Data Report*, 5 (6), 116–128.

Miller, A. P. (1986). Teleinformatics, transborder data flow and the emerging struggle for information: An introduction to the arrival of the new information age. *Columbia Journal of Law and Social Problems*, 20 (89), 89–144.

Masmoudi, M. (1984). The New World Information Order and direct broadcasting satellites. *Syracuse Journal of International Law and Commerce*, 8 (322), 322–342.

Modelski, G. (1972). *Principles of world politics*. New York: The Free Press.

Mowlana, H. (1986). *Global information and world communication, new frontiers in international relations*. New York: Longman.

Novotny, E. J. (1980). Transborder data flow and international law: A framework for policy-oriented inquiry. *Stanford Journal of International Law*, 16, 431–439.

Organization for Economic Cooperation and Development. (1980). *Guidelines governing the protection of privacy and transborder data flows of personal data*. Paris: OECD.

Robinson, P. (1980). Some economic dimensions of TDF. *Transnational Data Report*, 3 (3–4), 137–146.

Sardinas, J. L., Jr., & Sawyer, S. M. (1983, November). Transborder data flow regulation and multinational corporations. *Telecommunications*, pp. 59–62.

Turn, R. (Ed.). (1979). Transborder data flows: Concerns in privacy protection and free flow on information, 1, *Reported to the AFIPS Panel on Transborder Data Flows*, Washington, D.C.: American Federation of Information Processing Societies.

United Nations Centre on Transnational Corporations (UNCTC). (1981, September). *Transnational corporations and transborder data flow: An overview.* Seventh Session on UN Economic and Social Council Commission on Transnational Corporations, Geneva, Switzerland.

chapter 9
Values Underlying the Role of Telecommunications in Retailing

W. Wayne Talarzyk
Professor of Marketing
College of Business
The Ohio State University

Robert E. Widing, III
Assistant Professor
Weatherhead School of Management
Case Western Reserve University

INTRODUCTION

Satisfying consumer values is a necessary prerequisite to the success of any retailing system. This chapter examines new methods to enhance the distribution of information, products, and services in retailing, made possible by emerging telecommunications technologies. The discussion is designed to help explain the current state of telecommunications in retailing, evaluate the nature of telecommunication-based retailing innovations, and explore how well emerging retailing systems might satisfy consumer values.

The intended purpose of incorporating additional telecommunications in retailing is to more efficiently bridge time, space, and merchandising barriers that exist between the consumer and retailer, thereby increasing consumer access to, and use of, retail offerings. Indeed, many of the costs and constraints which regularly confront consumers engaging in product and information search and purchase may be sharply reduced or rendered moot by readily available technologies. This chapter will analyze a variety of telecommunications-based retailing methods that potentially serve to save consumers time, money, and effort, while simultaneously expanding the choice set and, perhaps, enhancing the quality of purchase decisions.

The first section of the chapter discusses a number of ways in which telecommunications-based technologies can enhance retailing operations. An examination of specific methods of communicating with, and marketing to, consumers through telecommunications is then conducted. This is followed by a delineation of the nature of technology-driven innovations, which reflects the need to search for consumer values that might be satisfied by coupling them with existing and emerging technologies. An evaluation of consumer values and how they might be affected by the emerging technologies is then presented. The chapter culminates in an analysis of selected public policy issues surrounding the expanded use of telecommunications in retailing.

OVERVIEW OF TECHNOLOGIES AVAILABLE TO RETAILERS

It is reasonable to surmise that the cultural determinants of a society include its values and technological capabilities. Values determine what a society desires and is willing to accept, whereas technology often determines the degree to which fundamental values can be served. Technology's role as a cultural determinant, however, has often been slighted in its importance and viewed as static in its capabilities. This is striking, given the unquestionably powerful impact technology has had on human progress (Nelson & Winter, 1977). Nevertheless, if an overriding goal of a society is to adapt to its environment or, conversely, adapt the environment to it, the single most powerful mechanism of adjustment is technological change (Rosenberg, 1986). This section provides an overview of some technologies available to retailers.

Retailers can choose from a growing array of technologies that potentially serve to both complement and substitute their current offerings. Such technologies offer greater control over internal operations and provide communications, information access, and transactional capabilities between sellers and their suppliers, customers, and service providers. In-home shopping services offer some of the greatest potential for change, but other technologies need to be considered, since, working together, they act to complement one another and enhance overall retailing opportunities.

Point of Sale Electronic Retailing

For example, Information Resources Inc. recently introduced the VideOcart, a computerized grocery shopping cart. Placed on the handle of a grocery cart is a 6-by-8 inch liquid-crystal display which is "connected" to the store's computer through an FM transmitter. Updated information is transmitted via satellite to individual stores.

The VideOcart has both market research and advertising capabilities. The research aspects include monitoring how long shoppers spend in each product area, which is helpful in analyzing product placement and store traffic decisions. The promotional applications include the triggering of ads as the cart moves past the product being advertised. Other uses include the advertisement of specials, video games that might be played in checkout lines, store maps to help locate products, and shopping recommendations ("Interactive Home Shopping," 1988).

A number of systems are expanding upon technologies similar to those used in ATM machines. For example, Advanced Promotion Technologies, a joint venture of Procter and Gamble, Donnelly Marketing, and CheckRobot Inc., has introduced a touchscreen system targeted for grocery stores, which offers a variety of options in addition to banking services. These include a store layout directory, receipts, coupons, sweepstake entries, weekly specials, and dispensing of instant refunds. This same venture has also introduced a self-service checkout system that allows the customer to personally use laser price scanners, eliminating the clerk function (Dillingham, 1988).

Other retailers offer in-store systems that range in capabilities from assistance in the selling process to enabling the buyer to consummate purchases through the system. Florsheim Shoe Company, for example, offers "electronic sales assistants" in over 200 locations called the Florsheim Express Shop. These outlets enable customers to view text, audio, and video information on over 425 shoe styles and conduct transactions (Sloan & Talarzyk, 1988).

Public Domain Videotex Systems

Public videotex systems have emerged in a number of localities, which are typically in high traffic sites and areas frequented by tourists. The systems usually incorporate easy-to-use touchscreen access and full color graphics. The latter are especially useful in providing diagrams for area locator maps, mall and store layouts, and merchandise displays. Some systems also allow users to print out desired information.

The Teleguide system, for example, is available in over 90 San Francisco Bay area sites and is fully financed through advertiser support. Listings include restaurant and hotel information (e.g., menus, prices, and locations on maps); information about tourist points of interest, conventions, and entertainment spots, mall and store directories; and transportation information, as well as general interest news and weather (Sloan & Talarzyk, 1988). Other cities/areas featuring similar systems include Toronto, Sacramento, Chicago, Cincinnati, Washington, D.C., Vancouver, and Central Florida.

Another application is exemplified by the Mall Information Center (MIC) at the Ingram Park Mall in San Antonio, Texas. This system uses an information center approach, with large-screen TV featuring short subjects, "Dee-jay" spots, and local advertising. Accompanying terminals provide mall, retailer, and product information.

Internal Organizational Support

A variety of electronic internal control and communications technologies are also available to the retailer. Laser price scanners help speed up the checkout process, eliminate the cost of marking/changing prices on individual items, and reduce pricing errors. The same technology can provide a wealth of sales and market research data at both the individual and group levels. In addition, information about inventory levels and reorder needs is made quickly available. The later is especially useful for centralized ordering at multiunit chains.

Rapidly diffusing check and credit verification systems and debit cards, which are often tied into the register system, are also highly useful cost containment innovations. Finally, medium-sized and large retailers are benefitting from the decreasing prices of telecommunications technologies such as FAX machines, teleconferencing, and data/voice transmission networks.

All of the above applications are part of the total computer, communication, and information support systems available to the retailer. However, perhaps the most innovative are the emerging electronic in-home retailing systems. The balance of the chapter is devoted to examining the types, applications, and value of these systems to both retailers and consumers. Specific in-home shopping methods are first examined which will provide a springboard for a discussion about how these technologies might serve to satisfy the values of certain consumer groups better than existing alternatives.

TELECOMMUNICATIONS AND IN-HOME SHOPPING

This section focuses on telecommunications-based in-home retailing applications. A wide range of in-home shopping options have emerged, differing in expense, complexity, consumer inertia, and "fit" with existing household hardware. The simpler applications represent extensions of traditional in-home shopping methods, while others require more radical changes in consumers' shopping behavior. Following is a discussion that groups shopping methods according to the degree of change, relative to current alternatives, which need be experienced in order for adoption to occur.

Telephone-Based Communications

The simplest telecommunications technologies to implement and evaluate involve those which augment traditional in-home shopping and require equipment the consumer already owns and is familiar with. These include the use of 800 and 900 numbers which consumers can use to place orders, make inquiries, register opinions and complaints, and obtain product-related information and assistance in product usage.

The seller-paid 800 numbers reduce time and "hassle" barriers in order placing, in comparison to traditional mail order, thereby enhancing selling and customer service efforts. The buyer-paid 900 numbers are not only used to market goods and services but are often the channel used to both deliver and simultaneously sell information and entertainment.

The 900 numbers applications entail the purchase of information or entertainment by the simple act of dialing the number. The services include the provision of information such as stock quotes, sport scores and gambling odds, recorded or live ribald messages, and many other theme-based messages. Another application includes paying for the right to register one's opinion in "polls."

Recorded information can also be provided at no charge to the user. Some newspaper publishers, such as the *Tulsa* (Oklahoma) *World,* provide over 40 recorded message services and depend upon advertiser revenues to support the system (Garneau, 1986). Many of the above-cited applications utilize what has come to be termed audiotex technology, which enables consumers with a touch-tone phone to interact with information bases. Following is a brief discussion of audiotex and some of its current and potential applications.

Audiotex

Audiotex, also termed *voice information services* and *interactive voice systems* (Greene, 1987), uses computerized or recorded voices to elicit responses and respond to caller requests. A system for airline reservations uses the typical menu format, where the voice asks if the caller is interested in either (a) flight status, (b) arrival/departure times, and so on. The caller selects an item by pressing the appropriate touch-tone telephone button. A computer reacts to the selection by listing additional sets of questions, which ultimately leads to a report of the desired information (Finnigan, 1986).

This same type of menu format can be used for a broad range of applications including personalized horoscopes, participating in surveys, pay-by-phone, catalog purchases, medical diagnoses, obtaining stock quotes and bank account information, and trouble-shooting product failures and recommending repair procedures. The latter represents the incorporation of diagnostic expert systems, where support numbers for computer software or hardware, copiers, consumer electronics, and the like, provide consumers access to 24-hour emergency repair help for common problems (Sheridan, 1987).

Early uses have been internal to a company, such as tellers interrogating a voice response system to obtain customer balances, or customer service representatives tracking inventories or customer shipments. More recent uses have expanded the user base to include customers who can interrogate the voice response system themselves. In essence, as a recent AT&T advertisement states, the 250 million phones in the United States are really computer terminals due to audiotex and related technologies. Eventually, voice recognition technology will enable consumers to use verbal responses instead of the push-button, multiple choice format.

Other related audiotex applications allow for the personalization of information through "dial out" capabilities. Users interested in news about a particular company, industry, or topic can have late-breaking information about the pre-selected topic immediately sent to their recorders (Finnigan, 1986; Widing & Talarzyk, 1983). Stockbrokers with changing recommendations, for instance, can immediately disseminate information to all clients. Similarly, other providers of time-sensitive information can tailor information to client needs. When combined with paging systems or "hot buttons" for critical topics, timely information access can be achieved.

Videotex

Videotex offers electronic transmission and computer-assisted analyses of information, which potentially enables consumers to break through existing barriers to information availability and processability. Videotex refers to systems that allow persons with a terminal or personal computer to access information and services from a central database(s), using phone or cable lines as the means of carriage (Widing & Talarzyk, 1983).

Beyond retailing applications, videotex offers a seemingly unlimited array of applications. These include, but are not limited to, banking and pay-by-phone services, boundless information resources from newspapers to the encyclopedia, electronic games, dating services, real-time communication with special interest groups, electronic bulletin boards, and electronic mail.

The retailing applications include the sale of tickets to an array of events; information and reservations for such services as airlines, restaurants, and hotels; the sale of regularly priced and deeply discounted brand name goods and services; and a host of other creative ideas. The latter include "11th hour pricing" of "sell it or lose it" offerings, such as cruise ship accommodations (deeply discounted to fill unused capacity shortly prior to sailing), and online product auctions.

The initial idea behind videotex, as envisioned by the innovator, Sam Fedida, in the late 1960s, was that this technology would enable mass consumer markets to obtain computer-listed information and place orders, as opposed to having a computer operator do so. Telephone information operators, retail order takers, reservation clerks, and the like would no longer need to interrogate the computer, since consumers could do it themselves (Wilkinson, 1980). This idea was expanded upon with additional services; however, the systems have yet to capture the imagination of mass consumer markets.

A number of well-publicized system failures, such as Knight-Ridder's Viewtron, Honeywell-Centel's Keyfax, the Times-Mirror's Gateway systems, have been discussed in the popular and business press at length. Some of the reasons for past failures have been hypothesized to include: market timing; little perceived value for the price; the need for dedicated terminals; limited advertiser support due to the small market; and problems such as the time and cost to "paint" graphics and the inability to present actual pictures of products, which is highly useful in stimulating product sales. (See Talarzyk & Widing, 1987, and Greenwald, this volume, for further analyses of such videotex failures.)

Less attention, however, has been paid to lower profile, yet relatively successful systems such as CompuServe, Genie, Dow Jones News/Retrieval, and The Source. These four systems have a combined subscriber base approaching 1 million, which is growing at better than 1% per month (Talarzyk & Widing, 1987).

Another large-scale system, Minitel in France, has experienced significant success and is in the early stages of entering the United States market under the name of U.S. Videotel (Nahon & Pointeau, 1987; Sloan & Talarzyk, 1988). In addition, IBM and Sear's long-awaited Trintex system, called Prodigy, has begun its 1988 rollout in San Francisco and Atlanta ("Trintex Pushes Videotex Ahead," 1988).

Two videotex retailing systems which have been available since the early 1980s include Comp-U-Card's pioneering Comp-U-Store and CompuServe's Electronic Mall. Comp-U-Store, for example, offers lowest available prices on a national basis and the capability to pay for and order models listed on the system. Price and attribute information are available for consumer electronics, durable goods, tableware, and a host of other products. In addition, the number of models listed in the database often range into the hundreds, far surpassing the typical number of models available in any retail store.

Despite a number of regional failures, therefore, other videotex and videotex-like systems continue to hold significant attraction and grow. To a great extent the early failures have financed the videotex learning curve, as opposed to ending it. Certainly, trial and error will continue to characterize these types of systems; however, the technologies are too compelling in promise for buyers and sellers not to eventually fill a major niche in the retailing arena.

Summary

The types of electronic retail technologies addressed here range in the degree to which they complement or substitute for traditional nonelectronic retailing. The 800/900-based numbers have been used in conjunction with a variety of electronic and traditional methods of retailing physical products, but largely in a complementary fashion with other media. Nevertheless, the 900 numbers can be used as a complete retail outlet for some products, primarily when the product itself is information. Similarly, audiotex systems, though more complex, are limited in their retailing capabilities, in the sense of a self-enclosed retailing system, and generally serve to complement other marketing efforts. For certain types of products, however, these too can constitute a complete retail service, especially in their interactive and "dial out" forms.

Retail activities in the videotex arena, such as those currently being undertaken by Sears, IBM, Comp-U-Store, and the Electronic Mall, present the greatest potential of a completely enclosed electronic shopping system. It needs to be highlighted, however, that even these videotex and videotex-like systems might often be used to complement, as opposed to strictly substitute for, more traditional retailing and shopping activities.

Comp-U-Store, for example, reports that a relatively small proportion of system inquiries result in a transaction. Instead, the systems are primarily used to obtain low price information which is useful in negotiating with local retailers. Further, one might speculate that sales made over the system were initiated by information obtained through retail store visits, recommendations made by product rating guides, and catalog browsing. The transaction made over the system represents only one step in the shopping process, perhaps taken due to product availability and dollar savings.

THE NATURE OF TELECOMMUNICATIONS INNOVATIONS IN RETAILING

The rate of diffusion and ultimate impact of telecommunications-based retailing, as with any technology, depends upon the fusion of technical feasibility and the presence of demand through the satisfaction of needs (Meyers & Marquis, 1969). As stated by Langrish, Gibbons, Evans, and Jevons (1972, p. 57), "perhaps the highest-level generalization that is safe to make about technological innovation is that it must involve synthesis of some kind of need with some kind of technical possibility."

The technologies discussed earlier have largely been the result of technical feasibility, or product driven, as opposed to market demand initially driving the development of technology to satisfy it. That is, videotex, audiotex, and even the telephone have largely been innovations originally stimulated by the supply-push model, as opposed to the demand-pull model, at the time of their inception.

In the in-home retailing arena, therefore, the key objective has been to discover needs that emerging technologies can satisfy better than existing alternatives. That is to say that, to a significant extent, in-home electronic retailing has been a solution in search of a problem. This is not, however, necessarily a problem. While the demand-pull model has traditionally been favored as the more important of the two models in characterizing successful innovations, this has been shown to misleadingly overlook the importance of supply-push innovations (Mowery & Rosenberg, 1979). Instead, understanding the product-driven nature of many retailing technologies should simply help indicate where analytical attention ought to be focused; that is, on gauging where market demand might exist for the various types of services and systems.

The supply-push model, for example, reflects numerous innovations which have often been borne from basic research, spin-offs, of research into other, originally unintended, applications, and accidental discoveries realized in the course of conducting research for unrelated purposes. In addition, the supply-push model also encompasses the leveraging of existing knowledge into new, albeit modest extentions. For example, the many and varied applications of polymer science, the spin-offs from knowledge gained from the space program into commercial products such as Teflon, the seredipitous lab discoveries such as Nutrasweet and the Post-It Pad adhesive, and the endless line extensions present in virtually every product class, are indicative of supply-push originated innovations.

Clearly, successful innovations occur which reflect both the demand-pull and supply-push models. Which of the two models initially drives an innovation is unimportant; indeed, the two approaches are often interactive and feed one another in a two-way causal stream. What is important, however, is that market demand and technological feasibility closely matchup for an innovation to be successful.

It is important to highlight, however, that satisfying values and, by extention, needs and wants, is not synonymous with market demand (Mowery & Rosenberg, 1979). *Market demand* is a precise term reserved for quantifying the relationship between price and quantity sold. The matching of needs and values with technical feasibility, in the absence of price and quantity information, can indicate likely failures but only suggest the *potential* of success. The lessons highlighted by the failure of a number of regional videotex systems attest to this proposition. Market demand did not exist for these systems, as people did not receive benefits sufficient to pay the price necessary to cover costs and provide an adequate return on investment to system operators.

People have many needs and value many things; however, these needs and values do not necessarily translate into market demand for a technology purporting to satisfy them. Additional factors such as price and other costs (e.g., learning) need to be incurred to use a system, which may not be compensated by the received benefits. Therefore, while an examination of the values satisfied and benefits offered by the systems is of interest, these other factors will play a large role in determining the rate of market acceptance and, consequently, the eventual shape of the services, systems, and industry.

A final issue related to the nature of telecommunications-based retailing innovations centers on industry life cycle concerns. Due to the early failures and slower than expected growth of, for example, videotex, a number of observers have reasonably questioned the viability of such systems ("Trintex Pushes Videotex Ahead," 1988; Saporito, 1987; George, 1987). However, it is necessary to remember that it is not unusual for new innovations to diffuse slowly or for early failures to occur.

Early product offerings generated by new technologies are often characterized by high cost, mediocre or limited applications and performance, and unreliability (e.g., copiers, FAX machines, ATMs, computers, VCRs, cordless and car phones, large screen TV, and so forth). This leads many industry observers to underestimate their eventual impact (Rosenberg, 1986).

Fortunately, few accepted technologies retain much semblance to the original versions. Instead, it is usually through major changes and refinements that seemingly unpromising new technologies eventually make an impact. In addition, innovations often depend upon the parallel diffusion and/or development of related technologies. For example, the penetration rates of certain videotex systems depend upon the diffusion rate of modem equipped personal computers. In summary, it is important to be mindful of the early life cycle stage in which many of the technologies discussed in this chapter reside, before dismissing them as having little promise.

VALUES AND ELECTRONIC RETAILING

This section will examine values in the context of how they might be positively addressed by the application of telecommunication technologies to retailing. Given that technical feasibility exists, a key to discerning potential demand is to examine those characteristics of electronic retailing systems that match up well with consumer values. It is necessary to keep in mind, however, the caveat stressed earlier about the difference between market demand and the satisfaction of needs, wants, and values. Satisfying values is a necessary but insufficient condition for market demand to exist.

List of Values

Values can be used in at least five ways for marketing purposes: market analysis and segmentation, product planning, promotional strategy, public policy, and positioning or repositioning a product (Vinson, Scott, & Lamont, 1977; Kahle, 1986). Kahle (1986) has identified and researched eight values termed as being fundamental to life's major roles. The values studied by Kahle (1984, 1985, 1986) serve as a useful starting point in evaluating the reviewed technologies.

The list of values (LOVs) can be thought of as building from Maslow's (1954) theory and is similar, though greatly reduced in number, to Rokeach's (1973) list of 18 terminal values. LOVs is, to a degree, a competitive formulation to SRI International's values and lifestyles (VALS) typology. This is due to both being usefully applied in marketing contexts, although Kahle provides research to support the greater predictability of LOVs over VALS on consumer behavior trends (Kahle, Beatty, & Homer, 1986). The eight values, ranked from most to least important according to how often each was cited by respondents as their primary value (Kahle, 1986), are briefly discussed below.

1. Self-respect. The most frequently cited value, which indicates that "Americans want to be at peace with themselves."
2. Security. This second most frequently cited value encompasses various aspects of security, including physical, psychological, and economic security.
3. Warm relationships with others. This value, which tended to be more frequently cited by women, highlights the importance placed on close social relationships.
4. A sense of accomplishment. This value was selected most often by people who have accomplished a lot.
5. Self-fulfillment. This value was described as a hybrid value that includes "individuality, accomplishment, and freedom." Those citing this value were quite similar to those who value "a sense of accomplishment."
6. Being well respected. The "external partner of self-respect," this value was cited most by those who felt they did not receive deserved respect.
7. A sense of belonging. The external partner of "warm relationships with others," which tended to be cited by women who devoted their lives to family.
8. Excitement, fun, and enjoyment. The least frequently cited primary value, which tended to be emphasized by young male pleasure seekers.

A problem with such high level abstractions, however, is that they tend to be removed from more concrete notions that are useful in evaluating specific dimensions of a technology or offering. The more concrete dimensions flow from, and feed into, one or more of the higher level abstractions. At a lower level of abstraction, there exist notions such as the consumer quests discussed below. Further, at a still more concrete level there are specific cost and performance attributes such as price, usage difficulty, features offered, etc.

Identifying values that might be satisfied by telecommunications-based retailing systems could be accomplished at the abstract level of fundamental values, by laddering each of the following consumer quests into one or more of the basic values. For example, the saving of time and money for other more preferred activities, simplifying and gaining increased control over one's life, enlarging product selections and enhancing product choices, and sharing information with other consumers could each be categorized as satisfying elements of security, self-fulfillment, accomplishment, a sense of belonging, etc.

The analysis which follows, however, will be conducted at the more concrete level of consumer quests and attributes sought, since this yields a more concise presentation. It is relevant, nevertheless, to emphasize that the fundamental values, such as those depicted by Kahle, serve as umbrella concepts under which the following consumer quests reside. Ultimately, telecommunications technologies must serve those demand-side values better than other options currently available to consumers, and in such a way that the benefits of usage exceed the costs.

Consumer Quests

The importance of in-home retailing and other forms of direct marketing can be seen by the volume of sales attributed to such activities. Over $150 billion is generated annually in response to direct marketing advertising ("DMA Statistics," 1986). Forty-five percent of all adults, and 50% of all female adults, purchase something through mail order in a given 12-month period (Simmons, 1985).

The reason for the magnitude and growth of in-home retailing can be at least partially explained by environmental change affecting consumers. Responses to these changes have been stated as trends or consumer quests by a number of observers (cf. Engle, Blackwell, & Miniard, 1986, chap. 11). The consumer quests addressed here include: (a) changing time perspectives, (b) instant gratification, (c) life simplification, (d) price/value and "do-it-yourself" orientations, and (e) the desire for self-tailored goods. Following is an examination of some of these quests in conjunction with telecommunications in retailing (for related discussions, see Urbany & Talarzyk, 1983).

Changing Time Perspectives

Changing time perspectives refers to the observation that, as money budgets increase, there is an increase in importance placed upon that portion of time budgets labeled as discretionary or leisure time (Voss & Blackwell, 1979). Saving people time, especially for shopping tasks that are "less than fun," is an increasingly important in-home retailing priority. For example, 81% of the sample in one survey cited saving time as one of the important reasons for mail order shopping, which was the most frequently cited reason of seven options (Webcraft Mail Order Survey, 1985; the second most important reason will be cited under "Self-tailored Purchases" discussed below).

Electronic in-home retailing should similarly benefit from time-poor consumers searching for time-saving methods. Major benefits offered by such in-home retailing include the capability to shop and order whenever the buyer wishes to, enabling transactions to be quickly consummated. The retailer also benefits, since there is less chance that an interested buyer will forget or delay purchasing.

Instant Gratification

Instant gratification is a time-related dimension but is somewhat different from simply saving time. This reflects the desire to have things now and with little inconvenience. Consumers, especially younger consumers, have little patience in regard to the postponement of gratification. Consumer research frequently indicates convenience as a major consideration by shoppers in their shopping behaviors. For example, one in-home shopping study reported "convenience" as being the second most important of six reasons for purchasing by mail, with 56.5% of the panel citing it as an important reason (Better Homes and Garden Consumer Panel, 1984; the most important reasons will be presented below under "Self-tailored Purchases").

All forms of electronic retailing feature convenience and instant gratification, or immediacy of response, as primary benefits. From obtaining sports scores to stock quotes from audiotex systems, to shopping and ordering over videotex systems, the convenient access and immediacy of feedback are compelling characteristics.

Although telecommunications can speed up information flows, there still exists the inherent time lag in physical delivery of ordered products. Instant gratification in terms of receiving the product remains problematic (English, 1985). This problem can be reduced through express delivery options for an added fee.

Life Simplification

Increasingly complex lives can be made simpler through the control and organization of information resources with audiotex and videotex systems. Having access to virtually an unlimited amount of information, ranging from product evaluation to articles provided by the national press and journals, is only one aspect of these systems. Perhaps more important is the capability to cope with these vast amounts of information through computer assistance.

The "dial out" capabilities of audiotex, described earlier, in which late-breaking information on a specified subject can be flagged and sent to the subscriber's tape player, is one example of simplifying the information environment. Another example is the computer assistance provided by videotex information providers, such as Comp-U-Card, in product search and evaluation. These represent extremely important information acquisition, organization, and screening tools for the user, whether the information is for business or personal use.

The ability to obtain relevant, comprehensive, and timely information, in a simple and very time efficient manner, is one of the great strengths of these systems. Computer-assisted acquisition and analysis of information promises to reduce consumer search and thinking costs, while simultaneously enhancing the effective amount of information processed, thereby enhancing the quality of decision making (Widing & Talarzyk, 1987, 1988).

Price/Value and the Do-It-Yourself Orientations

Increasing levels of education and consumer sophistication in evaluating alternatives have led to consumer demands for increased value from their purchase decisions (Engel, Blackwell, & Miniard, 1986). Perhaps one of the most important aspects of electronic retailing is the freedom they offer from local market constraints. National-based product and price searches enable consumers to transcend local markets. In addition, national product pools can be supplemented by third party information about the performance and quality of the alternative brands. Hence, greatly expanded comparison shopping opportunities, direct ordering and paying, and independent product evaluations combine to enhance choice while reducing the price paid.

Opportunities also exist for what has been termed *functional shiftability*, which is an extention of the "do-it-yourself" phenomena. This refers to the shifting of what have been retailer, wholesaler, or other intermediary's functions either back to the manufacturer or forward to the consumer. Manufacturers and consumers can be directly connected for searching, ordering, and paying, thereby replacing the retailer or others who have traditionally performed these functions (cf. Sauer, Young, & Talarzyk, 1988).

Electronic retailing offers the potential of spanning the gap separating manufacturers and consumers, at the expense of traditional intermediaries. The payoff comes in both wider selection due to the opening up of the national marketplace to the individual, as well as cost savings attributable to enhanced comparison shopping and economies gained by bypassing intermediaries. While the intermediaries' functions cannot be eliminated, electronic retailing systems allow for the consumer and manufacturer to more easily assume them. Hence, the potential gain in both price and value to the consumer.

Self-Tailored Purchase

This category extends the above but, due to its ramifications, deserves individual attention. Surveys of mail order customers have highlighted the premium placed on a wide selection. Two surveys were cited earlier, one that rated saving time as the foremost reason (of seven options) and the other rating convenience as the second most important (of six options). The second most important reason in the first survey was "selection is better" ("Webcraft Mail Order Survey," 1985). The most important in the second survey was "can't find items elsewhere" (Better Homes and Gardens Consumer Panel, 1984). These surveys point out the very critical role that broader product selection and availability plays in in-home retailing.

Electronic retailing presents the opportunity for the consumer not only to expand the available choice set of products from which to choose, but to interact with the production process. This capability, coined *prosumerism* (Toffler, 1982), allows for the consumer to dictate product specifications as the product is manufactured.

Emerging telecommunication and manufacturing technologies enable the buyer to again communicate directly with the seller. This allows for the creation of self-tailored goods, while retaining some of the economies of mass production. Hence, prosumerism should enhance the range of choice and result in a closer match between buyers needs and sellers' offerings. What have been called *splintering markets*, due to a fundamental desire for self-tailored goods and the improving capability to provide them, will become increasingly prevalent. Prosumerism represents a key, and likely dramatic step, in affecting the shape of distribution channels and, of course, the ability for consumers to "self-segment" themselves.

Summary

Numerous changes are posed to retailing structures by emerging developments in telecommunications. Certainly, these changes will likely occur gradually; however, a meaningful impact has already been felt by a number of electronic retailing applications. These include the impressive growth of computerized discount stock brokerage services, the success of some television shopping shows, the numerous audiotex and recorded message services, 800-number-supported direct marketing efforts, and the unspectacular but steady growth of certain videotex shopping services. The latter, over time, may prove to create the most dramatic changes in retailing methods.

Although in-home electronic retailing will not supplant in-store retailing, it will likely continue to substitute for, and change a number of, retail functions. Changes in marketshare will also be accounted for by the array of in-store electronic retailing methods and telecommunications support services.

The fundamental reason is that they serve, or potentially serve, to provide benefits that consumers value, including convenience, time-saving effort reduction, greater control over information resources, more and more relevant product information, a wider choice of alternatives, and the saving of money.

A great deal of experimentation will continue to take place. Learning and refining the systems by trial and error will mark most efforts. It is, therefore, important that retailers closely follow emerging innovations, examine how they might be positively or negatively affected by the new retailing forms, understand how the systems fit in with consumers' needs and values, exploit opportunities as they emerge, and prepare themselves for change prior to the occurrence of such shifts in consumer shopping behavior.

RETAILING, TELECOMMUNICATIONS, AND PUBLIC POLICY ISSUES

The examination of consumer quests and values, and the capability of the telecommunications technologies to satisfy them, was necessary to obtain some understanding of the viability of the systems. Assuming that these systems do eventually diffuse widely into the marketplace, certain public policy issues related to electronic retailing become relevant. These issues include system dependency, privacy, security, information equity, consumer decision making, and economic displacement.

System Dependency

System dependency is troublesome to the extent that retailers and service providers increasingly depend upon electronic distribution networks and hardware for controlling operations. "The computer is down" has become an increasingly familiar remark leading to irate customers who are delayed in accessing their bank accounts or records, placing orders with airlines and retailers, or obtaining needed data from information systems. ATM networks and branch bank offices depending on centralized databases, credit verification services, 800-number reliance, scanner and electronic register dependence for transactions and inventory control, and computerized records for innumerable organizations are just a few examples of vulnerable dependence.

As society becomes increasingly reliant upon electronic-based information systems, we become more vulnerable to system shutdowns. Although some technology fixes exist, such as purchasing redundant computer systems, phone or cable lines, and power supplies, businesses without them are in risk of having to simply shut down operations. It is key for policy makers to understand the vulnerability of many critical services to system breakdowns, in order to formulate measures geared toward reducing the damage created by these eventualities.

Privacy

Marketers, wishing to better understand their customers and profit from this knowledge, possess an array of electronic workhorses for obtaining market information. This information is not only at the group level, but the individual level as well. For examples, expert systems used to analyze scanner data from retail stores, when combined with cards identifying the individual shopper, can develop rather comprehensive product purchase records. This information can then be used in follow-up promotions sent directly to the consumer.

While rather benign in some contexts, much information gathered from customer interactions with videotex and audiotex systems is potentially more troublesome. These systems are quite capable of tracking purchases, information accessed, and communications. Given that some purchases and information obtained might be very sensitive to privacy concerns, and that seemingly personal information could be used in market research for further sales or sold to other interested parties, some reasonable question exists about the propriety of compiling and maintaining individual user profiles. Policies concerning the appropriate use of such information need to be more fully developed.

Security

A related issue to privacy and dependency is information security. Beyond maintaining appropriate account records and security of confidential information, there exists the problems of electronic theft and vandalism. One banker stated in regard to security shortcomings, "it's a problem we'll have to learn to live with . . . (but) if we waited to develop the perfect system, it would be 50 years before in-home banking becomes a reality" (Widing & Talarzyk, 1982).

While industry shields and safeguards are continually being developed, stories about "infected" databases and the theft or destruction of information are not irregular reading. As the systems are further developed, firms and individuals are increasingly likely to be exposed to security problems. In addition, an awareness of the magnitude of existing security shortcomings needs to be made clear. The reported incidents most likely grossly understate the actual number and extent of security violations, largely due to a hesitancy of firms to publicly acknowledge that their databases have been compromised. It seems desirable for policies concerning appropriate safeguards, as well as reasonable reporting requirements when databases have been violated, to be developed.

Information Equity

Profiles of videotex system subscribers consistently reveal the typical user to have higher than average incomes, job status, and education or, as some describe them, to be "upscale" (Widing & Talarzyk, 1987). These are the types of people who already acquire, use, and benefit from greater amounts of shopping information, thereby leading to their paying lower prices and making more informed purchase decisions.

The retailing innovations, as with other information and communications technologies, therefore, will likely most benefit those who, in a relative sense, stand to gain the least from them. An information elite may emerge, leading to a widening gap between the information rich and poor (Tydeman, Lupinski, Alder, Nyhan, & Zwinpfer, 1982; Widing & Talarzyk, 1982). This is compounded at the international level, as the gap among information-rich and -poor nations could similarly be affected.

Because of equity considerations, it has been suggested that resources be expanded in attempts to bridge the gap. The French government, for example, has been deeply involved in subsidizing their Mintel system, thereby expanding both small retailer and consumer access. Sweden, in response to similar equity concerns, took the rather drastic step of barring product rating information (Thorelli & Thorelli, 1977). Ironically, it has been well argued that the information seekers discipline manufacturers to the benefit of all consumers (Thorelli, Becker, & Englenow, 1975). Attempts to circumscribe information access, therefore, could lead to a lessening of product quality, which would adversely affect all consumers. Furthermore, it is quite possible that easier access to information, made possible by telecommunications, might actually enhance the "democratization" of information, as opposed to resulting in its increased concentration.

Consumer Decision Making

A primary goal of consumer policy, for both private and public initiatives, has been to enhance consumer decision making through the provision of product-related information (Thorelli & Thorelli, 1977; Beales, Mazis, Salop, & Staelin, 1981; Mazis, Staelin, Beales, & Salop, 1981; Maynes, Mergan, Weston, & Duncan, 1977). The tailoring of information, as discussed earlier, through applications such as "dial out" audiotex and the computer assistance offered by videotex information providers, promises to both hasten and improve product decision making. In addition, systems incorporating computer-assisted decision aids offer the potential to simultaneously enhance two key information providing standards, information availability and processability (Russo, Staelin, Nolan, Russell, & Metcalf, 1986; Widing & Talarzyk, 1988).

This enables what has come to be thought of as an inherent tradeoff to be overcome, namely, the amount of information provided and the ability to effectively process it (Talarzyk & Widing, 1988). Hence, circumscribed information provision policies, justified due to past research indicating an unwillingness or inability of users to acquire and analyze product information, need to be rethought.

Economic Displacement

New technologies not only create new capabilities and opportunities, but also displace old ways of doing business. Electronic cash registers went from 3% of the market in 1971 to 100% in 1978, displacing mechanical cash register manufacturers and their work forces (Craig, 1986). The same holds true for countless other economic displacements attributable to technology dating from the industrial revolution to the present. Relatively recently, technological change has led to the displacement of a number of jobs in manufacturing, railroads, natural resources (e.g., coal, forestry, and oil), and agriculture. Yet, since the mid-1970s to the mid-1980s, which were relatively high unemployment years, the number of employed persons increased from 80 million to 100 million (Rosenberg, 1986).

In the in-home retailing area, it was the inauguration of Parcel Post in 1913, hailed by the postmaster general of the day as the "greatest and most immediate success of any venture in the country," which "sealed the doom of the rural merchant" (Boorstin, 1973). Despite the branding of those who used the new retail methods as "traitors to the community," many rural merchants were swept aside by the convenience, price, and selection offered by mail order, which was greatly enhanced by the new delivery system.

As with most innovations a great deal of resistance can be expected by those who are likely to be negatively affected by emerging retail innovations. National marketplace access to a virtually unlimited variety of often discounted goods and services will create a new form of competition for local retailers. Videotex, as the most discontinuous innovation of those discussed in this chapter, represents the greatest long-term threat to the local market share currently held by retailers.

It is important for retailers to keep abreast of technological developments to determine how they might use them and, perhaps more importantly, avoid the negative effects of displacement. From a policy standpoint the question remains what, if anything, should be done. It seems clear that economic dislocation, within certain retail sectors, will occur. However, as Rosenberg (1986) writes, the short term dislocations due to technological change are apparently swamped by its positive effects on employment. Policy issues such as the smoothing of transitions, advance notification of shutdowns to employees, and the protection of affected business will, nevertheless, likely dominate debates.

CONCLUSION

The telecommunications technologies examined in this chapter are viewed as holding promise in satisfying consumer wants, needs, and values. The applications and variations of the various forms, as illustrated by the many creative initiatives cited above, seem to be limited only by the imagination and, for the more complex systems, the difficulty in gauging market demand. It is important to highlight, however, the very limited perspective one can have in trying to forecast the ultimate impact of the retailing technologies, not to mention unforeseen superceding innovations and the inevitable ripple effects on numerous economic and social issues. Nevertheless, some tentative concluding observations can be made.

The 800/900 telephone numbers and most audiotex applications are simpler to evaluate than the videotex applications. This is due to the former simply being extentions of current ways of doing business, as opposed to requiring a high degree of change. They are not terribly complex or difficult to use and understand, are dependent upon equipment that most consumers are familiar with and already own, and are relatively inexpensive to supply and use. These more continuous innovations and their applications will likely experience fairly quick acceptance and growth in the marketplace.

Videotex, however, represents a more discontinuous innovation. This technology also enables many other services to be offered in addition to retailing ones. Therefore, the future of the retailing applications are at least partly determined by the acceptance of other elements of the total offering. It appears safe to state that videotex will experience significantly more trial and error, as well as failures, over the foreseeable future. Although technological feasibility certainly exists, identifying valued services and successful methods of implementation are still being undertaken. Whether the benefits provided by the currently developing systems exceed the requisite costs to purchase and use the systems are less than clear. Certainly this is the case in regard to "mass market" implications.

When evaluating any new technology, however, it is key to remember that the final form rarely resembles the initial offering. Predictions of marketplace failure, based on the failure of early entrants, therefore, are very premature. Videotex offers applications that are compelling, since they are technically feasible and in consonance with market values. Over time, and with system revisions and price declines, videotex seems to offer too many *potential* cost and performance benefits to not make a mass market impact.

In summary, it seems more reasonable to conclude that the questions about telecommunications in retailing ought to be more of a "when and how," as opposed to "if," nature. It is important for both retailers and policy makers to monitor developments and prepare for the expanding role of telecommunications in retailing. For retailers the implications of not getting involved include losing marketshare to both in-store and in-home electronic retailing systems. For policy makers, the implications are broader. These range from questions concerning the possible regulation of information, the use of individual purchase histories, support for research on system usage and effects and, as with other governments, possible involvement in system development.

REFERENCES

Beales, H., Mazis, M., Salop, S., & Staelin, R. (1981, June). Consumer search and public policy. *Journal of Consumer Research*, pp. 11–22.

Better Homes and Garden Consumer Panel. (1984). *DMA Fact Book*. New York: Direct Marketing Association.

Boorstin, D. (1973). *The Americans: The democratic experience*. New York: Random House.

Craig, R. (1986). Seeking strategic advantage with technology?—Focus on customer value. *Long Range Planning, 19*, 50–56.

Dillingham, S. (1988, February 15). More Automation for Grocery Stores. *Insight*, p. 41.

DMA Statistics on Sales and Direct Response Advertising Expenditures. (1986). *DMA Fact Book*. New York: Direct Marketing Association.

Engel, J., Blackwell, R., & Miniard, P. (1986). *Consumer behavior*. New York: Holt, Rinehart and Winston.

English, W. (1985, Summer). The impact of electronic technology upon the marketing channel. *Journal of the Academy of Marketing Science*, pp. 57–71.

Finnigan, P. (1986, November). Audiotex: The telephone as data-access equipment. *Data Communications*, pp. 141–153.

Garneau, G. (1986, December). Telephone information services. *New Tech*, pp. 22–23.

George, R. J. (1987). In-home electronic shopping: Disappointing past, uncertain future. *Journal of Consumer Marketing*, 4(4), pp. 47–56.

Greene, R., (1987, December). Much more than horoscopes. *Audiotex Now*, pp. 7–8.

Interactive home shopping spurred by research. (1988, January 4). *Marketing News*, p. 40.

Kahle, L. (1984). *Attitudes and social adaptation: A person-situation interaction approach.* Oxford, England: Pergamon.

Kahle, L. (1985, Winter). Social values in the eighties. *Psychology and Marketing*, pp. 231–237.

Kahle, L. (1986, April). The nine nations of North America and the value basis of geographic segmentation. *Journal of Marketing*, pp. 37–47.

Kahle, L., Beatty, S., & Homer, P. (1986, December). Alternative measurement approaches to consumer values: The list of values (LOV) and values and life style (VALS). *Journal of Consumer Research*, pp. 405–409.

Langrish, J., Gibbons, M., Evans, W. G., & Jevons, F. R. (1972). *Wealth from Knowledge: A study of innovation in industry.* New York: Halsted/John Wiley.

Maslow, A. (1954). *Motivation and personality.* New York: Harper.

Maynes, E., Mergan, J., Weston, V., & Duncan, G. (1977, Summer). The Local Consumer information System: An Institution to Be? *The Journal of Consumer Affairs*, pp. 17–33.

Mazis, M. (1981, Winter). A framework for evaluating consumer information regulation. *Journal of Marketing*, p. 45.

Meyers, S., & Marquis, D. G. (1969). *Successful industrial innovations.* Washington, DC: National Science Foundation.

Mowery, D., & Rosenberg, N. (1979). The influence of market demand upon innovation: A critical review of some recent empirical studies. *Research Policy*, 8, 102–153.

Nahon, G., & Pointeau, E. (1987, January). Minitel Videotex in France: What we have learned. *Direct Marketing*, pp. 64–71.

Nelson, R., & Winter, S. (1977). In search of useful theory of innovation. *Research Policy*, 6, 36–76.

Rokeach, M. (1973). *The nature of human values.* New York: Free Press.

Rosenberg, N. (1986). The impact of technological innovation: A historical view. In R. Landau & N. Rosenberg (Eds.), *Positive sum strategy* (pp. 17–31). Washington, DC: National Academy Press.

Russo, J. E., Staelin, R., Nolan, C. A., Russell, G., & Metcalf, B. L. (1986, June). Nutrition information in the supermarket. *Journal of Consumer Research*, pp. 48–70.

Saporito, B. (1987, September 28). Are IBM and Sears crazy? or canny? *Fortune*, pp. 74–80.

Sauer, P., Young, M., & Talarzyk, W. W. (1988, April). *Functional Shiftability: Impact of New Technologies on Marketing.* (Ohio State University Working Paper Series).

Simmons Market Research Bureau. (1985). 1985 Study of Media and Markets. *DMA Fact Book.* New York: Direct Marketing Association.

Sloan, H., & Talarzyk, W. W. (1988, April). *Videotex project reviews VI.* (Ohio State University Working Paper Series).

Sheridan, J. (1987, June). Turning phones into terminals. *Industry Week*, pp. 58–59.

Talarzyk, W. W., & Widing, R. (1987). Videotex: Are We Having Fun Yet. In R. R. Dholakia & C. Surprenant (Eds.), *Marketing and Telecommunications in the Information Era: Strategies and Opportunities* (pp. 49–62). Providence, RI: University of Rhode Island.

Thorelli, H., & Thorelli, S. (1977). *Consumer information systems and consumer policy,* Cambridge, MA: Ballinger.

Thorelli, H., Becker, H., & Engledow, J. (1975). *The information seekers.* Cambridge, MA: Ballinger.

Toffler, A. (1980). *The third wave.* New York: William Morrow.

Trintex pushes videotex ahead. (1988, January 5). *USA Today,* p. 7B.

Tydeman, J., Lupinski, H., Adler, R., Nyhan, M., & Zwimpfer, L. (1982). *Teletext and videotex in the United States: Market potential, technology, public policy issues.* New York: Mcgraw Hill.

Urbany, J., & Talarzyk, W. W. (1983, Fall). Videotex: Implications for retailing. *Journal of Retailing,* pp. 36–92.

Vinson, D., Scott, J., & Lamont, L. (1977). The role of personal values in marketing and consumer behavior. *Journal of Marketing, 41,* 33–50.

Voss, J., & Blackwell, R. A. (1979). The Role of Time Resources. In O. C. Ferrell, S. Brown, & C. Lamb (Eds.). *Conceptual and theoretical developments in marketing* (pp. 296–311). Chicago: American Marketing Association.

Webcraft Mail Order Survey. (1985). *DMA Fact Book.* New York: Direct Marketing Association.

Widing, R., & Talarzyk, W. W. (1982, January). *Introduction to and issues with videotex: Implications for marketing.* (Ohio State University Working Paper Series).

Widing, R., & Talarzyk, W. W. (1983, Spring). Videotex: Where it stands. *American Council on Consumer Interests Proceedings.*

Widing, R., & Talarzyk, W. W. (1987, April). Methods for enhancing information search using videotex and related technologies. *American Council on Consumer Interests Proceedings.*

Widing, R., & Talarzyk, W. W. (1988, July). *Electronic information systems and consumer information provision policy.* (Ohio State University Working Paper Series.

Wilkinson, M. (1980). Viewdata: The Prestel systems. In E. Sigel (Ed.), *Videotext.* White Plains, NY: Knowledge Publications Inc.

chapter 10
The Consumer Videotex Market: Has It Reached Its Potential?

Marilyn Greenwald
Assistant Professor of Journalism
E. W. Scripps School of Journalism
Ohio University

An article published in 1983 in *Business Week* magazine ("A Surprising Eagerness," 1983) was not unlike hundreds of others in professional, scientific and popular publications around the world: videotex, it said, is the wave of the future, the savior of time, energy, and, eventually, money. And, the article predicted, within a decade, computer terminals using the videotex service would be common fixtures in most homes.

Since 1983, however, much has happened in the consumer videotex market—most of it not good, at least from the standpoint of those companies and organizations that market the service and manufacture the computers. The interactive video system that was predicted to have revolutionized the home office, journalism, and even the entertainment and travel industries, in the eyes of many, had stumbled. User projections for home systems had fallen flat. And, during the winter of 1986, two major companies that spent millions of dollars on videotex trials announced they were discontinuing their service. Knight-Ridder Inc. and Times Mirror Co., two publishing giants, waved the white flag. Perhaps home systems were not the wave of the future, they concluded (Arlen, 1986). Other videotex pioneers apparently felt the same way. As Talarzyk wrote, "Today most early regional videotex systems have been withdrawn from the market and the optimistic industry experts have retreated from their rosy prognostications. . . . Some industry observers are wondering if there really is a market for videotex and if so is the market of sufficient size to justify new entrants into the field" (Talarzyk & Widing, 1987)?

What are the reasons for the decline of videotex in the consumer market? It would seem that the technology is in place and works effectively and efficiently. But apparently the simple efficiency of a new technology is not enough to ensure its success, even in a competitive marketplace; a "human" element also enters into the picture. When it comes to some types of communication information, consumers sometimes are content to maintain the status quo; that is, fear of the technology combined with a reluctance to alter some living patterns has hindered the growth of videotex use in the consumer market.

It seems relatively clear that one of the problems of videotex is an apparent fear it instills on the part of prospective users. As is the case in many new technologies, consumer users are intimated by the product itself and their inability to immediately use it. In the case of videotex, apparently, this fear has never been completely overcome. Another seems to be the reluctance on the part of some private companies and public agencies to get into and stay in the market and to rely on any acceptance of change by consumers. Still another issue concerns turf protection by newspaper editors and publishers who feel the technology is a sort of double-edged sword. Can, they wonder, take a chance and dive into it, hoping for success, or they can wait and possibly be devoured by competitors?

Finally, an apparent "hands-off" policy by government regarding regulation gives the signal that the technology is not a serious one. Interestingly, the status of videotex is one that has divided many researchers and communication scholars. To some, the technology has been and continues to be successful. They cite its continually growing user lists as evidence of their argument. To others, videotex may be, as Talarzyk and Widing say (1987), "a solution in search of a problem"; that is, it is simply not, for most home users, an effective way of getting and sending information. The "solution-in-search-of-a-problem" advocates cite the discontinuation of many videotex projects as support for their argument.

Although the technology for videotex is readily available, even a willingness by consumers to use it does not necessarily guarantee its success. The reluctance by private industry to develop the technology further, and by the government to regulate it, are subtly changing the public's desire for it. Even if the technology appears to be efficient and caters to all of the values of modern society (it is cost effective, quick, simple and, in general, efficient), a simple consumer demand apparently is not enough to fuel its success.

Although predictions about the market vary, Tydeman, Lupinski, Adler, Nyhan, and Zwimpfer predicted in 1982 that, by the end of this century, 40% of households will have videotex services. They base this projection on the fact that it took television 16 years to penetrate 90% of all households. (It is interesting to note that, when projections for use of many new technologies are made, their growth is inevitably compared to that of television. This is done, perhaps, to illustrate that the public eventually embraced that strange new medium after some initial resistance. But, of course, the comparison of videotex and television is not valid. When television was introduced, users were getting an entirely new visual medium that brought the world into their homes, unlike anything they had ever seen. With videotex, consumers are simply getting another form of a medium that is familiar to them—they already have, for instance, newspapers, libraries, banking facilities and the like.)

And, as late as 1985, Alber projected four "phases" for the success of videotex: from 1973 until 1983 would be "evaluation and development," when developers would research what people wanted in a videotex system; from 1984 to 1987 would be "introduction to public services," when the marketing emphasis would be on consumer awareness of the technology; from 1988 until 1995 will be "rapid growth," characterized by "popular acceptance of videotex" and proliferation of in-house systems, with marketing emphasis shifting from educating consumers to selling specific products; finally, from 1996 to 2000 would be "maturity and stabilization," with a stress on "severe competition among firms in the industry." The services provided by the technology would (comprise) "a fundamental part of the way individuals live" (Alber, 1985).

Of course, the jury is still out on these projections. But, if anything, the use of videotex seems to lag behind Alber's projections. No one can say what will happen to the technology in the 1990s, but by 1988, it appeared that it had not achieved the "rapid growth" phase.

No matter what happens to the technology at the end of the 20th century, it is perhaps ironic that many of the same values that encouraged the growth of videotex in the early '80s in the United States contributed to its decline later in the decade. The convenience of the technology, ease of use, relative low price and its wide reach into business, information, and the entertainment industries would seem to make it ideal for the consumer. *Business Week* in 1983 reported the results of a new study by a management consultant firm that expected videotex to be the next "major home appliance." The article quoted users in a videotex trial praising the technology's versatility ("It could save a lot of time with chores like banking and shopping," one user said) and its ease of use.

But users may soon have discovered that the very qualities that make videotex so attractive can also be negative. For instance, the reader of the electronic newspaper can get news from across the globe instantaneously; but the reader has found, in many cases, that he or she wants the advertisements and graphics that a newspaper offers. While the user has instant access to his or her bank accounts, grocery prices, and airline schedules, he or she may have realized that others can also invade one's privacy. Here the interactive nature of the medium hurts the security of its users.

Perhaps the user does not want a machine and all those who have access to its data base to know the airline he or she prefers, the brand of groceries preferred, and, of course, the amount of money in bank accounts. Here, the value of privacy and security in the knowledge that one's life is not an open book are held higher than that of convenience. While efficiency and ease are top priorities for most Americans today, they clearly are not the only priorities.

Whose role is it to regulate the content and the providers of such technology?

Who must come up with public policy guidelines to establish a course of action in dealing with the economic and social issues of videotex? How will a terminal in every home and access to information from around the world blur the traditional "roles" of home and office? Will the technology further split a society that is deeply divided among the rich and the poor?

These apparently are some of the value questions raised by videotex users and providers. While the use of videotex by the consumer is far from nil, its progress has perhaps been hampered by issues and questions that may never have arisen during its birth. Consumers have indicated that they are not willing to pay the price—in more than just dollars—for a bit of convenience. As Talarzyk writes:

> Consumers are willing to pay for products and services which save them time, effort and money. The problem seems to be that there are not enough consumers that see the values of videotex exceeding the "costs" (time, effort and dollars) associated with using such systems . . . What are the unfulfilled needs of the marketplace and how can videotex best fill those needs compared to other alternatives currently available . . . Either the present perceived values of videotex will have to increase or present costs for the services will have to decrease. (Talarzyk & Widing, 1987)

The "cost" of videotex is not simply a financial one, as far as consumer values. They fear the technology and, unlike television, microwave ovens, video cassette recorders, and other examples of widely accepted new technologies, the consumer does not seem willing to believe that the services of videotex offset the fears.

THE TECHNOLOGY

To better understand some of the problems and concerns associated with videotex, one must first understand the technology itself. Talarzyk describes videotex as a "marriage" between computer and communications technologies. Usually, telephones in combination with computer modems are used to transmit data, although in some cases television cables are used. One can see immediately the regulation problems here; because of the form of transmission, should videotex be regulated by the same rules governing telephones? Or should rules applying to broadcasting be used? Or, perhaps like newspapers, no group should be sent up as "regulators." By 1987, the 700,000 subscribers to the three major national videotex services (CompuServe, Dow Jones, and The Source) have access to in-home shopping and banking services, obscuring even more the public-policy debate about videotex. Several agencies could enter the picture. It could be the Federal Trade Commission, since the question remains whether consumer protection laws apply to electronic sales of goods. It could be the

Federal Reserve, since the transfer of money is involved in videotex. It also could be the Federal Communications Commission, since equal-time rules that affect broadcasters could apply to videotex. Finally, it could involve the Copyright Tribunal, because the question of information ownership may arise. The ambiguous nature of these questions has not been fully clarified, leaving prospective videotex providers in a quandary by adding an additional element of risk.

The question remains then—just why has videotex failed to meet some of the glowing projections for growth of the early '80s? Talarzyk offers several reasons for the failure of regional videotex systems, including the facts that (a) consumers will not pay the costs of a dedicated terminal but instead would like more ways to use their personal computers, (b) the number of consumers ready for videotex is too small to support a regional system, (c) advertisers are not willing to pay requested rates to reach such a small market as videotex, (d) consumers are not willing to pay the extra costs of the graphics associated with a videotex system, and (e) many of the systems were too "news oriented."

If videotex were an end to itself and provided services that the user could not get elsewhere, perhaps more prospective users would pay the price for a dedicated terminal. In this case, the value of its efficiency is simply not great for the consumer. In this way, videotex developers may have, again, overestimated the importance of "efficiency" in the lives of consumers.

The growth and success of media organizations worldwide may indicate that today's consumers want to be informed about what is going on in the world. But this desire for information and its importance as a value is apparently not overwhelming. Prospective videotex users evidently believe they can get their news quickly and easily from numerous sources and do not necessarily need it from their computer terminals. This need to get the news as soon as it happens, i.e., the ability of some data bases to offer users news as it comes over the news wire, is clearly not a top priority.

Talarzyk's argument that advertisers may not be willing to pay for the graphics associated with the system is a "chicken–egg" argument. Clearly, if regional videotex systems did reach more people, advertisers would be willing to absorb the costs.

Finally, the importance of news to prospective videotex users has clearly been overrated by videotex or database developers. Consumers today are perhaps inundated by what they feel is up-to-date news in newspapers, television, and radio. They may feel they are getting all the news they need, and getting more, even faster, may not be a top priority for most Americans.

When Knight-Ridder announced in 1986 that it was abandoning its pilot videotex project, Viewtron, it stated in a news release that, despite steady growth in the number of subscribers, "usage numbers have not kept pace. . . . It is . . . clear the American public is not ready to suport a videotex service that would justify the continuing expense." The company said it had about 3,000 subscribers when it withdrew from the market in March 1986. Interestingly, Knight-Ridder, Times Mirror Co., and most other newspaper publishing companies entered videotex with the idea that the technology might some day eliminate the need for newspapers. As the Knight-Ridder news release pointed out, "It is now clear that videotex is not likely to be a threat to either newspaper advertising or readership in the foreseeable future" (Talarzyk & Widing, 1987).

FEAR AND THE ELECTRONIC NEWSPAPER

An interesting case study of the fear instilled by a new technology is the videotex project known as the electronic newspaper. In the late 1970s and early 1980s, many newspaper publishing companies like Knight-Ridder, fearing they would be left out on the cutting edge of a new technology, spent millions on videotex pilot projects which, in essence, provided "newspapers" over the computer or television screen. Readers could get "headlines" from a menu, and call the stories they wanted to read.

The videotex newspaper projects (most of which have been abandoned) illustrated two points: first, that readers like to browse through newspapers and even carry them with them from room to room, home to job, and the like. The portability of an actual newspaper, not to mention the fact that it allows readers to cut out and save coupons, photos, stories of interest and the like, is clearly important to consumers. Surely this "control" the reader has over the paper is important to him or her. Whether print-outs of stories, coupons, and other features of newspapers fulfill that need for control remains to be seen, but it is unlikely.

Of course, the first to cry foul with the advent of the electronic newspaper were the editors and writers themselves, who saw the new-fangled invention as opening a can of worms. First, they were worried about their jobs. Second, they were worried that the rights afforded them by the First Amendment could be in jeopardy. Third, they worried that they might suddenly face competition in the news business from those who might never before have entered the "newspaper" business.

Wicklein writes, "To some extent, First Amendment protections accorded newspapers give them a competitive advantage over broadcasting, which they see as a rival medium. If they do not get involved in videotex, they are not as likely to fight *its* First Amendment battles as well" (Wicklein, p. 144).

The concerns among journalists didn't stop there. In the April 1983 edition of *Quill,* the magazine of the Society of Professional Journalists, several stories indicated the fear and distaste journalists had for electronic newspapers and other new technologies. In that issue, one journalism instructor says that technology itself "outdistances terminology and thought. The difference can be more than semantic. It can be attitudinal" (Aumente, 1983, pp. 10–15). Still other journalism instructors and newspaper staff members write that all newspaper people must re-learn how to write with the electronic newspaper, since writing style is more terse and to the point. As one wrote, in *Videotex/Teletext*, "Letter formation and graphic are crude. . . . The viewer has to wait for a story of interest. . . . The crowding of the medium with conflicting messages is highly distracting" (Alber, 1985). Of course, those aren't the only problems with electronic newspapers, according to those in the newspaper business. Gerald Stone (1983), who teaches journalism, wrote in *Quill*, "Call it disinterest, laziness or just practicality. The news-reading public is not going to sit in front of a screen full of scrolling words."

Another value concern about the electronic newspaper not mentioned as openly by editors and publishers is the issue of the Fairness Doctrine. When deciding how far to go with videotex and the electronic newspaper, it is likely that publishers such as Knight-Ridder and Times Mirror Co. took into account whether that particular doctrine would apply to them. The Federal Communications Commission required that, under the Fairness Doctrine, broadcasters and cable operators devote a reasonable amount of time to convey controversial issues of public importance, and that coverage of those issues must be fair and present opposing points of view. Although there has been some debate on whether the Fairness Doctrine violates the First Amendment, the commission upheld it and reaffirmed it after a 1974 study. In 1988, the doctrine was rescinded, although the possibility that it may be reinstated remains.

But some researchers like Anthony Oettinger, director of the Program on Information Technologies and Public Policy at Harvard University, believe affirmation of the Fairness Doctrine was a way for government to eventually regulate the print media (Oettinger, 1988). Certainly, publishers saw this as a drawback to the promotion of the videotex and to its widespread success.

It is clear that the electronic newspaper was not highly promoted by those whose job it was to produce it, possibly contributing to disinterest among consumers. Most if not all of the pilot projects were not successful. Davenport wrote in 1987 that, indeed, it was the fear among newspaper publishers (those who have control over publications) that overstated the so-called demise of videotex, when it may be in fact alive and well. She writes: "Why is it there is not much coverage or positive news on the progress of videotex and why do editors think there are big problems associated with it even though it has almost a million subscribers and continues to climb? . . . Is it because upper management is worried about competition from videotex" (Davenport, 1987, p. 107). Davenport also notes that the failure of the Knight-Ridder and Times Mirror projects accounted for less than 3% of all videotex users.

Still, the fate of the electronic newspaper was not the only indication of the consumers' dampening enthusiasm for videotex. It suddenly dawned on many that, if, through the technology, they could invade the files of others, many could invade their files or private lives. For instance, one researcher of the British Prestel system suggested that consumers let their computers, through videotex, do their "browsing" for them; the consumers could let the system know they want to be aware of any new development in a certain science, for instance. Or they could go one step further and program it to let them be supplied with all items of a certain political nature. As Wicklein writes, "Here is where the trepidation is. Although you have entered your instructions into the computer, it is not the computer that is doing your idea-shopping for you, purely and objectively. It is another human being, making subjective judgments and loading the machine with items that . . . will appear to meet the criteria you have laid down" (Wicklein, 1981, p. 73).

In addition, Wicklein points out, questionnaires sent out by videotex services to learn of consumer needs could potentially backfire. The computer could know of individual preferences and emotional needs and (combining it with "profiles" of such things as past entertainment and airline ticket purchases) use the information to sell to businesses, political parties, or even the government. Furthermore, Wicklein points out, it would be easy for governmental agencies "legally" to get access to bank and other accounts. The list goes on. Credit ratings, as well as credit card information, could be available to anyone and updated daily (Wicklein, 1981, p. 97).

Of course, not all researchers into the new technologies agree that privacy is as much an issue as some make of it. Joshua Meyrowitz in No Sense of Place (1985), an examination of the effects of television on society, believes that any technology that gives the user access gives him or her power. Therefore, fear of loss of privacy is not justified. "The thing to fear is not the loss of privacy per se, but the nonreciprocal loss of privacy—that is, we lose the ability to monitor those who monitor us" (Meyrowitz, 1985, p. 322). Others might believe that, with the advanced technology that currently exists, the lives of most people are an open book as it is. That is, any bank employee could, with a bit of effort, learn the financial condition of any bank customer with existing technology, as could any employee of credit-checking companies. Some people may believe that Americans have already lost most of their privacy.

SOCIAL AND POLITICAL RAMIFICATIONS

Any discussion of a new technology would have to include its effects on society. Videotex is not different. If the line is blurred between our jobs and our home lives, this will dramatically effect society. In No Sense of Place, Meyrowitz argues that the "global village" impact of television, which brought the world into everyone's living room, had a pervasive effect on nearly all aspects of society. The same could be said for videotex, if it indeed some day reaches initial projections. As Tydeman, Lupinski, Adler, Nyhan, and Zwimpfer (1982, p. 256) write, spouses may be drawn to each other as much as for their ability to manipulate databases as their cooking or athletic talents.

One must also ask the effect the automated home on child rearing, women's rights, and even education. Tydeman et al. note that the "electronic cottage," as they call it, could free women to have traditional careers (with computers at home that link them to the world) and rear children. But teaching children by age may no longer be effective. Current educational systems segregate children by age and grades, while the computer makes no such distinction. The technology creates groups of people based on interests and skills, regardless of their ages.

How would this segregation by ability affect the education system? How would the lack of interaction of children of the same age within a classroom affect the socialization process? It would seem that, if indeed children were taught in groups based on ability rather than age, would indeed hinder the learning process—the education about personalities that children get from each other. Because this socialization is so subtle, its value may not be fully realized by parents.

As Tydeman et al. (1982, p. 265) note, the advent of videotex in the home could dramatically change American politics. For instance, public opinion, as well as information on referenda such as a new mass-transit schedule, could be garnered electronically instead of personally; that is, people would not have to go to public hearings or the polls to voice their opinions or get information. The same could be said for candidates themselves—with at least one drawback: "Citizens who lodge complaints may find that their electronic profile of past political involvement is exposed" (p. 266).

Of course, the immediate transmission of voting information into the home could have the same effect as exit polling and remain open to the same criticism; that is, if users on the East Coast see from West Coast results (via videotex) that one candidate is winning by a large margin, voters on the East Coast may decide not to waste their time going to the polls. Consequently, the technology could have a dramatic effect on the election process.

The point here is that widespread use of videotex in the home can have dramatic and rather instant ramifications. They are effects that are not quite as subtle or long-term as those of television and other new technologies.

In brief, the social and political ramifications of widespread home videotex usage could be great. But one has to ask if such questions arose in the minds of the users or providers who, in essence, rejected videotex. Did Knight-Ridder, for instance, decide it didn't like the changes in voting patterns that videotex might spawn, and was that a reason it discontinued its videotex trial? Certainly, one must deny this assumption. To Knight-Ridder and other videotex "experimenters," bottom-line financial reasons probably had top priority in their decision to discontinue the trial. But such social ramifications may have a secondary effect on the failure of videotex. That is, such social and political ramifications may have been on the minds of the prospective users of the service, contributing to their rejection of the systems.

ECONOMIC AND PUBLIC POLICY CONSIDERATIONS

As the history of videotex in the United States has indicated, a mere disinterest by consumers can lead to bigger problems that can stunt the growth of a new technology. Videotex may, in fact, soon become an orphan technology. Outside the United States, government agencies have played a major role in developing videotex systems (an issue to be discussed later). This has not, obviously, been the case in the United States. The withdrawal of many large publishing companies from the field, and the breakup of AT&T, a prospective player in the videotex market, raise questions about who will enter the arena. Obviously, the advantages of a government-aided videotex market (like those in Great Britain, France, and Sweden) are many to the consumer user, especially regarding cost. In the United States, however, private companies are encouraged to add charges to the services they provide for profit, in keeping with the spirit of competition. In fact, there have been few if any attempts by recent United States presidents to establish government agencies to encourage and support growth of new telecommunications technologies. Wicklein writes of a memo sent to clients of Kidder Peabody & Co., a New York brokerage firm: "In the U.S., there exists an obstacle to the success of a (videotex) service. We believe that countries in which videotex is likely to operate successfully and quickly are those in which the communication carrier has a virtual monopoly and is not barred from offering computer services" (Wicklein, 1981, p. 86). Apparently, Kidder Peabody did not mean a government monopoly, but a private one such as AT&T. And, of course, the government indirectly prompted AT&T to enter the videotex business; with the break-up of the corporation came the understanding that it could enter fields such as videotex.

Wicklein is quick to point out that, in addition to the "strings-tied" results of a government-aided system, it is clear the Pentagon would be "an influential silent partner . . . the Department of Defense already controls one-third of the electronic spectrum assigned to satellite and microwave communication" (Wicklein, 1981, p. 9).

Because of a reluctance on the part of the U.S. government to participate in the development and growth of videotex, advertising from private companies is likely to continue to play a role in its growth. Advertising revenues will likely subsidize the cost of the service and help cut costs to consumers. Again, if given the chance, advertising through videotex could be specifically tailored to certain audiences. It is naive to think, however, that a videotex system free of government intervention would be totally free of outside influence. Often the underwriters of programming on public television (which sometimes includes the federal government) can exert enormous control on programming. And the government, as television underwriter in France and Britain, also has control. In Britain, people who pay for television licenses often pressure the BBC for certain programming. And in France, where the government finances the television system, the party in power decides what will air.

PUBLIC POLICY AND REGULATION

The role (or lack thereof) of the U.S. government in establishing regulation of videotex probably contributed to the declining status of the technology in the United States. First, the nebulous role the government has taken in the regulation of the technology may have helped stunt its growth. Second, the questions remaining about who actually regulates videotex also have had a hand in hurting its growth.

Before one examines the public policy issues associated with videotex, one must first examine how some other countries have dealt with regulation and laws. One of the most well-known videotex systems in the world is Prestel, a commercial operation started in 1979 by the British Post Office. Developed in 1970, Prestel was initially known called a "Viewdata" system, which came to be a generic name for videotex. Prestel enabled users to select from 250,000 pages of textual material viewed on a television screen. The user connects a small adapter and key pad connecting the television to a computer through telephone lines.

The system came into being when the BPO was seeking ways to improve telephone usage, specifically during the evening hours. (The British Post Office once had a telecommunications arm.) Like many other systems, Prestel used a "tree" for searching a broad category. The tree led to several "branches" as more specificity was sought. As for regulation, Wicklein writes, "The BPO deliberately shied away from setting standards for what was being transmitted over its 'common carrier' system, saying these were the province of the IPs (information providers) themselves. The IPs, aware that government regulation of their advertising and promotional conduct was inevitable, got together to draw up their own code of conduct to forestall more government restrictions" (Wicklein, 1981, p. 69). But, Wicklein notes, questions arose as to how such regulation would be enforced.

It is perhaps ironic that France, known for a telecommunications system that lagged behind those of other developed nations, was also ahead of many countries in videotex. By 1980, after Prestel came into being, many nations, including France, Japan, Sweden, Canada, and others, began some videotex pilot projects. In the 1960s, France had only three million telephones in a country of 50 million people.

Interestingly, television and radio viewership was popular and encouraged by the government, while two-way communication by telephone was not (the broadcast media were controlled by the government's TDF, or Telediffusion de France).

As the demand for telephones grew, the TDF realized this was a potential for increased revenue and began research into videotex. The French government, however, planned to serve as its own information provider, despite controversy over whether the French people would accept such an arrangement. As part of the plan, 25,000 telephone subscribers were given electronic "telephone directories," which proved to be the focal point of the French videotex system. Each telephone was supplied with a black-and-white screen and a terminal. Subscribers merely typed the name they wanted and the telephone number appeared on the screen. Gradually, some additional information such as weather and other government-provided information became available. Eventually, the list of subscribers grew, and the French, with their electronic telephone directory, became frontrunners in the videotex field.

Unlike Britain and France's, the United States government has not been a driving force behind the development of videotex. This is due in part to the nature of the government, but also to deregulation. Where other countries have provided as little as subsidies and as much as overall management, the U.S. government has little if any involvement in the technology. Of course, the question remains—how much of a hand *should* the U.S. government have in the establishment of videotex? But, if it takes no role, the technology may never meet its potential. Mosco wrote in 1982 that the only direct government presence in a videotex system came through an agricultural information experiment, Project Green Thumb, when the Departments of Agriculture and Commerce attempted to bring weather, market, and production information to farmers over their television sets. Even that small experiment was open to criticism by those who felt it "illustrate(d) the danger of federal encroachment on private sector perogatives" (Mosco, 1982, p. 87).

Of course, the main barrier to regulation of videotex in the United States is simply defining what the technology is. As Tydeman et al. write, "Videotext (is) emerging in the United States in a policy environment that has a tradition of separating various types of communication media (i.e., broadcasting, point-to-point communication, newspaper)" (Tydeman et al., 1982, p. 171). Yet because videotex represents "the merger on a mass consumer basis of (regulated) communications and (nonregulated) computer technologies," it challenges policy distinctions (see Figure 10.1). In short, Tydeman et al. claim, the changes in technology have blurred the distinctions within telecommunication media and the distinction between nonelectronic and electronic media. Tydeman et al. go into detail about how several broad areas of government, including the First Amendment, the Federal Communications Act, and antitrust laws, impinge on videotex. Their observations are interesting; for instance, the First Amend-

Technology	Traditional Regulatory Assumptions	Challenging Factors
Newspaper	No government involvement; constitutionally guaranteed free press promotes multiplicity of voices	Electronic newspaper using telephone lines and cable TV systems currently being tested
Broadcasting	Locally based radio-TV with licensee serving as public trustee; limited spectrum must be regulated to ensure public interest	Multiplicity of video sources including distant signals on cable, pay programming, videocassettes, videodiscs
Telephone	Service provided on nondiscriminatory basis through common carriers; rate regulation; no content regulation	Increased competition for equipment, service
Computers	No direct government intervention; market-driven technology	Merging of computer and communications technologies
Mail	Government monopoly for first-class mail; uniform rates; cannot be used for illegal purposes; limited subsidy for distribution of books, newspapers and periodicals; universal service guaranteed	Electronic mail commercially available

From *Teletext and videotex in the U.S.*, by J. Tydeman, H. Lupinski, R. P. Adler, M. Nyhan, and L. Zwimpfer, 1982, New York: McGraw-Hill. Copyright © 1982 by McGraw-Hill. Reprinted by permission.

Figure 10.1.   Media Regulation Vehicles

ment, as interpreted by the U.S. Supreme Court, does not provide freedom of speech for advertising or false or libelous speech and has singled out radio and television as areas where limited frequency availability justifies government regulation. So, Tydeman et al. ask, will electronic newspapers have the same First Amendment protections as newspapers, or the more limited protection of the broadcast media? This blurring of the differences between the print and broadcast media brought on by videotex brings to light numerous questions. It certainly must have influenced the decision by large newspaper publishing companies to remove themselves from the videotex market.

In general, four options exist regarding public policy and its role in videotex regulation, Tydeman et al. believe. First is the possibility of no videotex regulation. Second is selective government regulation by application of laws to specific kinds of content (such as obscenity or advertising). Third is government application of different degrees of content regulation, depending on the medium of transmission; and fourth, self-regulation by information providers (of course, questions arise about enforcement here).

Technology	Rate Regulation	Content Regulation	Structural Regulation
Broadcasting	None	FCC: Fairness Doctrine, Equal time rules (under debate) Ban on some advertising; Requirement for local programming	FCC: Limit on no. of stations per owner; Limit on TV-newspaper cross-ownership
Telephone	Tariffs set by FCC and by state PUCs	No "mass media" services by AT&T	Common carrier rules on access; FCC: Computer inquires; Courts: Anti-trust suit; Congress: Comm. Act
Cable	Local franchise or state regulation	FCC: Broadcasting-type rules applied to origination cable-casting by cable operator; "Must carry rule" on local signals; Local franchises may require specific program services (e.g., local access)	FCC-imposed ban on cross-ownership of cable and TV station in same market; Ban on network ownership of cable; Ban on telephone co. ownership of cable in urban markets
Print	None	Case law on obscenity, libel	None

From *Teletext and videotex in the U.S.*, by J. Tydeman, H. Lupinski, R. P. Adler, M. Nyhan, and L. Zwimpfer, 1982, New York: McGraw-Hill. Copyright © 1982 by McGraw-Hill. Reprinted by permission.

Figure 10.2.   Comparison of Regulatory Frameworks for Videotex Distribution Networks

Even if public policy issues are identified, one must ask who is responsible for the content and quality of service. Content rules themselves apply to broadcast and cable, but not to newspapers, telephone service, or cable channels that are not controlled by a cable company (see Figure 10.2).

THE FUTURE

Despite some discouraging signs and its failure to meet many projections, the future of consumer use of videotex is not completely bleak. For instance, CompuServe, one of the nation's leading videotex companies, had in 1987 about 360,000 subscribers, and adds about 8,000 each year, according to figures from the company. In addition, changes in ownership of some large videotex companies may also have a positive effect on growth. Also, as more businesses use computers, the consumer may gradually overcome his or her fear of the key pad and of employing the use of one at home. Once the user is comfortable with the technology outside the home, he or she will be more open to using it at home.

But possible new problems exist. Currently under discussion by the Federal Communications Commission is a proposal to change the implementation of an access charge to time users of videotex. Currently, videotex users pay a fixed monthly access charge, similar to one charged long-distance telephone users. Under consideration is a proposal to charge *according to the hour*. According to Charles McCall (1988), chief executive officer of CompuServe, the FCC believes the change would be more equitable to telephone users. Needless to say, videotex companies and users oppose the proposal and are doing their part to let the federal government know. Such a proposal would undoubtedly raise the costs of using videotex and perhaps limit the extent of its use.

Still, the advent of videotex has raised some important and interesting questions regarding values and public policy in American society. Are, for instance, the values of efficiency and convenience, touted as so important in modern times, actually top priority for most Americans? Or are traditional values of security, socialization and human contact more important? And how much of a role do simple habit and custom play in the acceptance of a new technology? In this period of very rapid technological change, some clarity of values is essential, or the arena of public choice will be adversely affected.

REFERENCES

A surprising eagerness to sign up for videotex. (1983, April). *Business Week*, p. 108.

Alber, A. F. (1985). *Videotex/teletext*. New York: McGraw-Hill.

Arlen, G. (1986, December). High rollers with high hopes. *Channels*, pp. 86–87.

Aumente, J. (1983, April). Room at the bottom: Nobody knows the talents they'll need. *Quill*, pp. 10–15.

Davenport, L. D. (1987). *A coorientation analysis of newspaper editors' and readers' attitudes toward videotex, online news and databases: A study of perception and options*. Unpublished doctoral dissertation, Ohio University.

Meyrowitz, J. (1985). *No sense of place*. New York: Oxford University Press.

McCall, C. (1988, February 18). Chief Executive Office, CompuServe, in speech to Ohio State University students and faculty members.

Mosco, V. (1982). *Pushbutton fantasies: Critical perspectives on videotex and information technology*. Norwood, NJ: Ablex Publishing Corp.

Oettinger, A. (1988, January 14). *The future direction of telecommunications and information resources policy*. Speech delivered to students and faculty members at The Ohio State University.

Stone, G. C. (1983, April). Alternative to a videotext future. *Quill*, pp. 17–21.
Talarzyk, W. W., & Widing, R. E. III. (1987, June). *Videotex: Are we having fun yet?* (Working Paper Series, College of Business, The Ohio State University).
Tydeman, J., Lupinski, H., Adler, R. P., Nyhan, M., & Zwimpfer, L. (1982). *Teletext and videotex in the U.S.* New York: McGraw-Hill.
Wicklein, J. (1981). *Electronic nightmare.* Boston: Beacon Press.

III
Telecommunications and Space

chapter 11
Communications in Space: Economics and Public Policy Issues

Molly K. Macauley
Fellow
Resources for the Future
Washington, D.C.

INTRODUCTION

Outer space as a vantage point for acquiring and transmitting information represents a significant and growing technological development in the global information economy. Yet contentious international debate surrounds both the technology of space and the information communicated by the technology. Nowhere is this contention more pronounced than in the cases of satellite communications and remote sensing (space-based observing of, e.g., weather, oceanic behavior, vegetation patterns, atmospheric pollution). In satellite communications, these issues have come to the fore prominently in international dispute over access to the geostationary orbit, a precise location in space that is required by most communications satellites. In recent years, the orbit has become increasingly congested, leading to concern over possible limits on its availability to all who wish to use it. In contrast to this problem, which is fundamentally a concern about technology access, remote sensing involves questions of flows of information. Remote sensing information is about a country's physical and natural resources, and in some cases, economic activity. Accordingly, issues of debate are over who should have access to and pay for this information.

These dual contentions over technology and information are not unique to space telecommunications policy; they prevail in the general context of telecommunications policy, as exemplified in discussion about universal telephone service in the wake of AT&T divestiture, or concerns over intellectual property rights to information. Space is a broader concern for four reasons, however: (a) the role of government, which must mediate the allocation of resources between private and civil government activities in space; (b) space-based or space-related scientific research, which competes for resources with commercial

185

interests; (c) the inherently global nature of space-based telecommunications technologies, which can create tension between national and international, and developed and developing country interests; and (d) the use of space technology to demonstrate national technological development or to symbolize national prestige, which introduces conflict between those who would treat space as a means to an end, and those who view space as an end in itself.

These concerns represent significant economic, social, and political stakes. This chapter outlines frameworks for considering the fundamentally economic issues, but the analysis addresses a broad socioeconomic and primarily non-market setting. It suggests explicit hypotheses about the interdependencies or system effects characterizing the various competing parties, and implicit incentives and signals that bring forth responses in this setting. Problems of collective management figure prominently in the analysis, particularly in light of the wide sectoral concerns between private and government, and national and international interest. The framework seeks to highlight divergences, and their implications, between private and social ends, and to point out alternative uses of society's resources to mediate scarce means and competing ends.

This chapter also suggests a set of values that underlie the ultimate use and interpretation of policy towards space and terrestrial communications technology. Figure 11.1 notes some of these values; several, such as environmental quality or national autonomy, arise from consideration of space technologies and reflect the vantage point of space or its inherently global nature. These values may not have appeared on a similar list compiled for, say, terrestrial technology alone. The shrinkage of time and space is evident, manifested in awareness, convenience, or privacy. The disbenefits of the technologies should not go unnoted, however; contrast the *inconvenience* or *invasion of privacy* that might attend interruption by an unwanted telephone call. How these values play a role

---

Awareness (of news, etc.)
Convenience (time)
Autonomy (of an individual, business entity, or nation)
Fairness
Distribution of wealth
Prestige
Leadership
Efficient use of resources
Standard of living
Privacy
Quality of the environment
Culture
Technological progress

Figure 11.1. Values Inherent in Public Policy for Space and Terrestrial Communications

will become clear in the sections that follow, and in the summary of the concluding section.

The remainder of the chapter proceeds in three sections. The second section
describes the technology of satellite communications, the nature of public debate about this technology, and the economic perspectives that seek to clarify
and inform debate. The third section follows this course in the case of remote
sensing. The last section briefly considers the positive (or descriptive) analyses
contained here in relation to the normative (or value) judgments implicit in
public discussion.

SATELLITE COMMUNICATIONS: THE POLICY ISSUES IN ALLOCATING THE GEOSTATIONARY ORBIT

Much of the debate concerning satellite communications technology takes place
during international deliberations over a particular nonmarketed resource required for the technology—the geostationary orbit. Debate over access to the
orbit may in fact reflect concern over access to satellite technology per se, or
over opportunities to use the technology to demonstrate national technological
progress or acquire prestige. For reasons outlined below, however, resource allocation at the fundamental level of orbital access is a key determinant in
reconciliation of the other concerns.

The geostationary orbit is a particular location in space that allows satellites
to orbit as if they are stationary over a point on earth, thereby simultaneously
interconnecting a large geographic region. Geostationary satellites transmit the
bulk of cable television programming to local cable distributors, and share in the
transmission of long-distance telephone calls, live sports and news broadcasts,
the text of newspapers and weekly magazines, and intracorporate data. In regions where terrain or the sparse distribution of population makes terrestrial
communications technologies prohibitively costly, satellites assume an even
larger role in providing basic communications infrastructure. Both developed
and developing countries have thus long been concerned with obtaining access
to the orbit.

The orbit is some 170,000 miles long, but limiting its use are signal interference between adjacent satellites, and the geographic earth coverage uniquely
associated with each location along the orbit. For instance, Figure 11.2 indicates
regions of coverage by country given various orbital locations (referenced by
their cartographic longitudinal degree). Orbital positions of use to the United
States, for example, are overlapped by positions of interest to Mexico, Central
America, and Latin America.

Procedures have been established to maintain distance between satellites to
mitigate intersatellite interference. Intersatellite separation reduces the number
of locations available to a region, however. Policy to handle the unique and

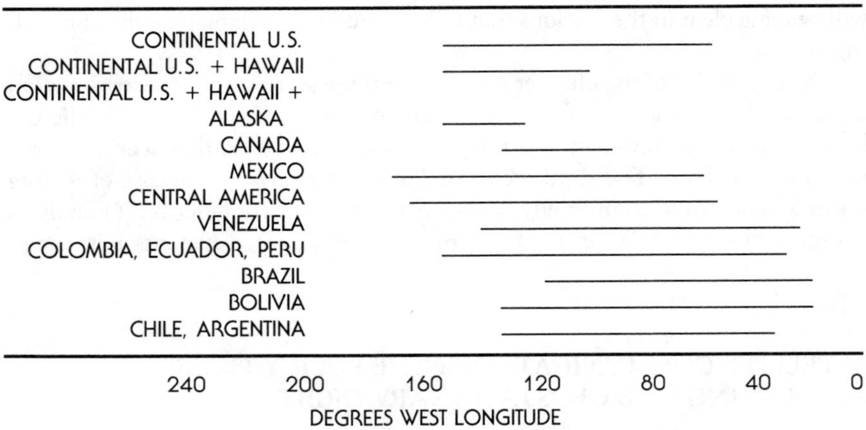

CONTINENTAL U.S.	
CONTINENTAL U.S. + HAWAII	
CONTINENTAL U.S. + HAWAII + ALASKA	
CANADA	
MEXICO	
CENTRAL AMERICA	
VENEZUELA	
COLOMBIA, ECUADOR, PERU	
BRAZIL	
BOLIVIA	
CHILE, ARGENTINA	

```
   240        200        160        120        80        40        0
                    DEGREES WEST LONGITUDE
```

From *Orbit-spectrum between the fixed-Satellite and Broadcasting-Satellite Services with applications to 12 GHz domestic systems*, by E. E. Reinhart, 1974, Santa Monica, CA: Rand Corporation. Copyright © 1974. Reprinted by permission.

Figure 11.2.   Usable Orbital Arcs

scarce locational advantages has tended to be that of "first come, first served." In recent years, however, the demand for orbital locations by both domestic U.S. industry and international users has greatly increased, rendering the non-market allocation and regulatory process administratively complex and highly political.[1] Accordingly, consideration of the orbit as the common heritage of all mankind has led, after much debate, to the imposition of stringent technical standards for satellite and earth-station operation to accommodate large demand for the orbit.[2] While the standards may thus provide orbital access for, say, developing countries, compliance with the standards has at the same time tended to increase the investment and operating costs of satellite technology.[3] In

---

[1] The relevant administrative institutions are, at the international level, the International Telecommunications Union, under whose aegis ongoing deliberations for orbit access policy are next scheduled to resume at the Space World Administrative Radio Conference in mid 1988, and in the United States, the Federal Communications Commission and the National Telecommunications and Information Administration.

[2] See, for example, Christol (1986) or Goroves (1985) for discussion of the principle of the "common heritage" as it has come to be reflected in treaties governing space. More precisely, it is outer space, broadly defined, that is regarded as the common heritage of mankind; different views have been expressed about the orbit. For instance, the 1976 Bogota Declaration states that the orbit (which is located over the equator) should be governed by equatorial countries.

[3] For estimates of orbit values, and the effects of operating standards on values, see Macauley (1987).

addition, a large amount of excess operating capacity (estimated at about 40%) now prevails in the domestic U.S. industry,[4] reflecting in part the regulation-induced response of domestic firms to increase their operating capacity per orbit allocation.

This discussion suggests several purposes of an economic investigation of the orbit. First, in the broader context of satellite insurance and launch policy, the need to make "hardware-intensive" use of the orbit to comply with new operating parameters may in turn result in complex satellites that are more costly to insure or launch, or that mismatch operating capacity and actual demand for satellite service. These effects impose burdens on both developed and developing countries (particularly with respect to realization of values listed in Figure 11.1). To mitigate the burden on developing countries, standards may be revoked or relaxed, yet might not alternatives (such as that of making allocations rentable) be preferable? In this case, developing countries would be able to lease allocations to any country or firm having the technology to use it, perhaps for the ten or so years that the satellite would be functional. Lessor nations could, if they wished, use the rental income in the interim to develop internal telecommunications networks and later occupy orbit allocations.

Another purpose of an economic framework for analysis is to understand how orbit regulation may affect the pace and direction of research and development in satellite technology. Here, the concern focuses primarily on domestic U.S. and European industry, where R&D has been channeled quite successfully into methods that economize on the orbit. Government-sponsored research on advanced communications technology satellites in the U.S. has also been justified in large part by a perceived need to develop methods that use the orbit more intensively. To the extent orbit scarcity is an artifact of regulation, however, industry (and ultimately, consumers of satellite services) would benefit from the opportunity to explore alternative areas of technical innovation, and to redirect government spending towards these areas.

REMOTE SENSING: TECHNOLOGY, INCENTIVES, AND INDUSTRY ORGANIZATION[*]

In the case of remote sensing, access to the technology itself is in many respects of lesser policy importance than in the case of communications satellite technology. This is not to say that ownership and operation of remote-sensing systems fail to demonstrate technical prowess; in fact, recent U.S. policy debate has included argument over the desirability of fostering U.S. leadership in this

---

[4] See Day (1987).

[*] Extensive conversations with Mike Toman underlie much of the discussion of remote sensing.

regard (an example of the values noted in Figure 11.1). Yet actual access to information about natural and physical resources continues to be of primary interest, and in this regard two distinct but related economic problems underlie much of the debate over remote-sensing information. One problem is that information or knowledge differs from conventional economic commodities in that it is expensive to produce and cheap to reproduce. Thus, it basically cannot be appropriated; it can be used on a nonexclusive basis by anyone who obtains it. The end result tends to be underinvestment in the generation of information. This problem is lessened when information is useful to specific firms or economic sectors (for instance, market product research or detailed demographic information). But the second problem arises because remote sensing at least in part conveys information about public goods, such as natural and physical resources, oceanographic and atmospheric research, and environmental monitoring, thereby conferring diffuse benefits to a large number of people and compounding the "free rider" problem. The problem increases in scope when the information pertains say, to global environmental monitoring. At the international level, arrangements thus need to address not only the assignment of rights to an intangible good (information), but the assignment of rights to information about which a country may be sensitive (for example, as a winner or loser in its perception of environmental responsibility—recall Figure 11.1).

Since its inception in the mid-1960s, civilian earth observation from space has significantly contributed to human knowledge and well-being, in areas ranging from measuring the earth's resource base to environmental monitoring and land-use planning. The technology uses satellite-borne sensors to identify natural and physical resources based on differences in their reflectance properties.

Contentious debate has always surrounded remote sensing in light of the appropriability and public goods problems. Specific subjects of dispute have thus included development and management of the technology, the nature and amount of data collected, the users to whom data are made available, and who pays for these activities. The debate also is heightened in large part by conflicts (some perceived, some inherent) among government and private objectives. From a governmental perspective, civilian remote sensing has been pursued with such aims as fostering national technological leadership, augmenting scientific data for basic research of earth processes, expanding engineering knowledge (of space operations, spacecraft, space sensors, and sensing techniques), and providing "public goods" such as weather information. In contrast to these social goals, numerous private sector objectives have been articulated, such as supplying commercial fishing and ship routing data, and offering private enterprise new business opportunities in the collection, processing, and dissemination of data. Most recently, the media have begun to use remote-sensing information to supplement and illustrate news analyses (recall photographs of the fire at the Chernobyl nuclear plant, the views of a Soviet laser base, and pictures of Iran's Silkworm missile sites in the Persian Gulf). Finally, some applications (perhaps most notably land and resource mapping) mix public and private benefit.

While public and private sector objectives need not always conflict, in the history of remote-sensing programs they seem to have been irreconcilable. The conflict also has fueled divisiveness among governmental agencies, between research and commercial interests, and between domestic and international entities. It appears to have intensified further with difficulties encountered in implementing the Land Remote-Sensing Commercialization Act of 1984, which sought to transfer operation of the U.S. civil land remote-sensing systems to the private sector. Problems have also arisen in the development of the recently formed EOSAT Corporation (the private sector firm now responsible for the U.S. civil system), and international concerns have been posed by the successful development and operation of the European SPOT earth observation system.

Moreover, new and potential uses of remote-sensing information are increasingly challenging existing public policy for information technologies. For example, the media use noted earlier involves information about which countries can be particularly sensitive. The following exchange highlights one aspect of the dilemma, in this case between media and national security interests. The exchange was between retired Air Force Major General Jack E. Thomas, a consultant with the Defense Department, and ABC's Mark Brender, who chairs a media task force on remote sensing for the Radio and Television News Directors Association: "Suppose," Thomas said, "that 300 hostages are being held somewhere. The media get a picture of an aircraft carrier . . . covered with helicopters. A rescue effort is about to be launched. Publication of that picture will destroy that mission." Brender's reply was "Surely the Soviets know that carrier is there. They know where it is . . . When we have to, when national security and the lives of Americans are at risk, there is tremendous press restraint."[5] To take another example, in the U.S. the Constitutional protection afforded by the first and fourth amendments has also been an issue in use of the technology, in the case of press freedom (the first amendment) and freedom from unreasonable searches (the fourth amendment). Fourth amendment issues surfaced when, for instance, the Environmental Protection Agency (EPA) used aircraft photography of a corporate site to monitor and enforce compliance with environmental legislation.[6]

The resolution of these issues has critical implications not only for the design and implementation of national remote-sensing policy, but also for the evaluation of worldwide initiatives. These initiatives include proposals for international "consortia" to organize remote-sensing activities, and the possible development in the 1990s of a global environmental monitoring system incorporating

---

[5] Reported in Rheem (1987, p. 1). See also Glaser and Brender (1986) for discussion of media use of remote sensing.

[6] For excellent discussion, see Russell, Harrington, and Vaughan (1986). The Supreme Court in 1986 held that the taking of aerial photographs of an industrial plant from navigable airspace is not a search prohibited by the fourth amendment.

space technology. [7] In these cases, economic considerations (including access to data, cost-sharing among participants in systems operation, and arrangements for data collection and processing) will loom large as nations individually access their ability and willingness to cooperate in multilateral initiatives. In addition, how countries internally reconcile government and private sector roles will assume international significance.

In addition to the appropriability and public goods problems, several other fundamentally economic issues are at the heart of remote-sensing policy. In particular, characteristics of the technology critically affect the incentives facing individual suppliers of the information and, in turn, the organization of the industry. These characteristics also significantly affect the desirability of alternative institutional arrangements and the usefulness of international participation in remote-sensing initiatives. These concerns have lacked investigation to date, leaving a gap in knowledge that has impeded remote-sensing policy.

Foremost among technological concerns with significant economic implications are:

Economies of scale and scope. Earth observation is a "multiproduct technology," in that it yields many types of data ranging from observations of climate or vegetation to evidence of physical resources and human alterations of them. This characteristic leads to complex questions about the extent of "economies of scope" (that is, the extent of cost savings from jointly producing different information outputs, as opposed to separate earth observing systems for each output) as well as to questions about economies of scale. Emerging innovation has tended to focus on the design of platforms able to support multiple sensors, in contrast to existing practices that use separate spacecraft for individual tasks, but the choice of technology calls for explicit evaluation of scale and scope economies. Platforms may also facilitate the processing of data in space, thus reducing the need to merge information on the ground and perhaps in turn lowering the cost of information. The use of platforms, however, also requires arranging for resource allocation and cost sharing among disparate government agencies and between governmental and commercial participants.

Economies of scope and scale will affect the efficiency of decisions that allocate production activities and cost between the public and private sectors, whether these decisions result from government policy or from commercial development without government involvement. By the same token, these economic factors bear strongly on the desirability of international participation in remote-sensing efforts, and on the design and effectiveness of institutional arrangements for such efforts. For example, if sufficiently large cost savings were attainable from international sharing arrangements that exploit scale and scope

---

[7] See McElroy, Clapp, and Hock (1987) and references cited therein for extensive discussion of international organization of remote-sensing activities.

economies, such cost savings could outweigh alternatives, such as separately owned and operated systems, that are otherwise favored because they better demonstrate technological leadership. The future of global habitability or international environmental monitoring programs also figures prominently as a factor in the viability of multilateral agreements in remote sensing.[8] Individual countries may support or discourage remote sensing depending on whether they perceive themselves to be victims or culprits in pollution abatement. Finally, just as a "clean technical interface" has been found to be demanded by countries possessing technologies, to prohibit technology flows, so too is a "clean economic interface" likely to be important in arranging for equitable data access and cost sharing.

Vertical integration. Remote sensing also requires a large amount of "value-added" processing of the raw transmitted data. This aspect of the technology poses further questions concerning economies of scope in the collection and processing of data, with consequences for the efficient vertical organization of remote-sensing activity.

Economically "sunk" costs. In addition, the technology requires significant investments in both ground facilities and spacecraft. The investments might be viewed as discouraging entry into remote sensing by new entities (whether they are national governments or private enterprises). However, for entry to be limited, the outlays must be not only large, but also "sunk" in an economic sense—that is, not salvageable or transferable to other uses. While launch costs obviously are sunk, ground equipment or sensors on spacecraft in low earth orbit may be reconfigurable (at least in principle) via the space shuttle or space station, and irretrievable investment at the value-added stage may be insignificant. Thus, assessing the height of entry barriers and evaluating their effect on incentives for national and private sector investment require detailed economic consideration of these technology aspects.

Product differentiation and uneconomic duplication of facilities. Capital investment and entry barriers are related to product differentiation. The nature of remote-sensing information can be changed by altering the mix and resolution of sensors or the spacecraft's orbital parameters. Under what circumstances will the right amount of diversity in data be provided to satisfy individual demand (including that from the public and private sectors, and from national and international interests), without excessive duplication of investment and effort?

Aspects of demand. Demand considerations also include problems of "exclusion," which arise when consumption cannot be limited to a specified group (for

---

[8] The signing in September 1987 of the International Convention for the Protection of the Ozone Layer (essentially, an agreement to reduce use of chlorofluorocarbons) is one harbinger of emerging multilateral willingness to address environmental concerns, and of the potential for remote-sensing technology to assume increasing prominence in monitoring and arbitrating such concerns.

example, those willing to pay for the service). For example, who should have first access to data? Should any nation have the right to censor the data to remove sensitive material? Should any nation have exclusive rights to the data? Should a nation have the right to levy a charge based on the amount of territory sensed or other criteria?[9] Investments in excluding consumers may be a critical part of the technology for commercial goods, but not for remote-sensing information deemed to be in the public domain.[10] The potential for asymmetry of exclusion (for example, a nation cannot physically prevent remote sensing of its territory but could be prevented from accessing information without payment) further complicates the analysis.[11] Liability issues might also be expected to arise for bad data, conflicts of interest, or an inadequate scientific base. The case of weather information is an example, where in a recent lawsuit the National Weather Service was alleged to have failed to issue a correct forecast, the absence of which was said to have led to the death of several fishermen.

Such issues reflect features of information as an economic good which lead to problems of discerning consumers' preferences and in turn, designing payment schemes that elicit the mix and quantity of remote-sensing data best able to satisfy all public and private interests. The scale of demand is a critical determinant of industry scale, with scale in turn influencing how the industry may be most effectively organized and managed at the domestic and international levels.

THE VANTAGE POINT OF ECONOMICS

The descriptive assessment of resource allocation in the case of the geostationary orbit, and problems of the economics of information in the case of remote sensing, assert predictions about behavior, but generally shirk assumption of ethical views about the desirability of certain kinds of behavior. That is, the economic issues as outlined above constitute a descriptive analysis that seeks to set forth "what is" as a basis for improving public policy for space-based commu-

---

[9] These queries are among concerns noted by McLucas (1985).

[10] As Glaser and Brender note (1986, p. 66): "International law has tended to favor open dissemination of information from remote-sensing satellites. It has long been accepted that remote sensing by ships outside territorial waters or by aircraft not in sovereign airspace is legitimate, as is the dissemination of information acquired by these means. The prevailing legal view has been that sensing from the sovereignty-free area of space is likewise legal. However, a number of states are arguing that although remote sensing is done in space, the data are intended for use on earth, where state sovereignty should prevail."

[11] There are ways to assign intellectual property rights or otherwise protect private providers—as have been practiced in the microchip and software industries. This aspect of remote-sensing technology has its counterpart in other telecommunications technologies, although in the case of remote sensing, the publicness of the information is key.

nications technologies. They do not purport to set forth a normative or proscriptive approach to "what ought to be."

Without going well beyond the scope of this chapter, some uneasiness over the distinction between positive and normative analysis persists after decades of contention. J. N. Keynes, Friedman, and Blaug are among those who argue that the distinction not only *can* be made but is rather sharp; reactions and counterarguments include those of Koopmans, Rotwein, and Lange.[12] As Blaug notes, "Even means and ends cannot be distinguished without specific value premises" (p. 707), and this assertion must necessarily be kept in mind in thinking through the controversies surrounding space technology. For instance, it underlies discussion of whether space activity should be undertaken for national leadership or for economic development (recall Figure 11.1), and whether international consortia are feasible for organizing remote sensing. In this regard, perhaps positive analysis contributes most significantly by evaluating the trade-offs involved in these choices—for example, under what conditions can the pursuit of space activity for national public benefit comport with the robust development of commercial space activity? Positive analysis can also make explicit the consequences of the relative weights that decisionmakers may be assigning to policy alternatives. For instance, the tension imparted by the joint supply of public and private goods in remote sensing is predictable and to some extent measurable by positive analysis.

Policy disagreement may also often not be over what is desirable (say, economic equality) but how to achieve it. In this case, positive analysis sheds light on alternative courses of action that may be obscured in usual debate. To illustrate by analogy with satellite technology and dispute over the geostationary orbit, the problem of maintaining environmental quality (like space, "the common heritage of mankind") also resulted in contentious public debates and then pollution control standards based solely on technical feasibility, without regard for the cost of compliance. Some economic incentives have now successfully replaced technical requirements in controlling air pollution. Analogously, there is some best match of orbit allocations and technical controls on orbit use. Economic analysis can help identify that match and facilitate more informed positions by national representatives during international deliberations over access to the orbit. Alternatively, such analysis could play a role in the centrally managed orbit planning process now underway in international forums.[13]

---

[12] See citations in Blaug (1978, pp. 1–9 and chap. 16).

[13] Quantitative information, and computer models to evaluate it, are increasingly being used in this process. This information primarily includes technical operating parameters, but economic data could easily be appended. Moreover, ITU reports are now making frequent reference to the need to take economic considerations into account in orbit policy. See Federal Communications Commission, Fifth Notice of Inquiry, Docket 80-741, May 4, 1987, pp. 15–22; and references in International Telecommunications Union, "Report to the Second Session of the Conference," World Administrative Radio Conference on the Use of the Geostationary-Satellite Orbit and the Planning of Space Services Utilizing It, First Session, Geneva, 1985, section 3.4.3.5.1.

Economic analysis can also elevate the substance of what at times is highly emotive debate. It has done so, for example, by suggesting an objective framework for considering environmental resources. Similarly and irrespective of whether marketlike mechanisms are ever embraced to organize use of the orbit, economic analysis can serve to inform (but need not necessarily take sides in) debate over the orbit. The perspective can uncover a range of unintended consequences that may implicitly value the orbit at less than its worth, and suggest policy changes that might render the orbit more effectively as well as more judiciously managed.

In the case of remote sensing, policy development appears headed toward serious consideration of the forming of international arrangements to organize the industry. Yet in pursuing international collaboration in remote sensing, attention needs to be given to reviewing the problems that have been encountered in other international endeavors that share fundamental aspects of space activity—such as resource management, information production and dissemination, the conduct of basic research, or the application of advanced technology. Candidates for study include the International Energy Agency, the World Meteorological Organization, the International Institute for Applied Systems Analysis, international cooperation in public health, and the law-of-the-sea negotiations. Each of these arrangements has met varying degrees of both success and difficulty in mediating perceived national self-interest, monitoring or sustaining collaboration, and ensuring that collective gains exceed those of individual action. Timely considerations of these arrangements for remote sensing will well serve policy development in this area.

REFERENCES

Blaug, M. (1978). *Economic theory in retrospect.* Cambridge, England: Cambridge University Press.

Christol, C. Q. (1986). The search for a stable regulatory framework. In D. Demac (Ed.), *Tracing new orbits: Cooperation and competition in global satellite development* (pp. 3–18). New York: Columbia University Press.

Day, S. M. D. (1987). Challenges ahead for the communications satellite industry. In M. Macauley (Ed.), *Economics and technology in U.S. space policy* (pp. 57–68). (Proceedings of a symposium sponsored by Resources for the Future and the National Academy of Engineering, Washington, DC, June 24–25, 1986.)

Glaser, P. E., & Brender, M. E. (1986). The First Amendment in space: News gathering from satellites. *Issues in Science and Technology, 3* (1), 60–67.

Goroves, S. (1985). Expectations in space law: A peek into the future. *Journal of International Affairs, 39* (1), 167–174.

Krasnow, R. (1986). Climate forecasting: Assessing an informational technology. In R. Krasnow (Ed.), *Policy aspects of climate forecasting.* (Proceedings of a seminar sponsored by the National Center for Food and Agricultural Policy, Resources for the Future, Washington, DC, March 4, 1986.)

Macauley, M. K. (1987). The contribution of a partnership between economics and technology. In M. Macauley (Ed.), *Economics and technology in U.S. space policy* (pp. 3–21). (Proceedings of a symposium sponsored by Resources for the Future and the National Academy of Engineering, Washington, DC, June 24–25, 1986).

Macauley, M. K., & Portney, P. R. (1984, July/August). Property rights in orbit. *Regulation.*

McElroy, J. H., Clapp, J., & Hock, J. C. (1987). Earth observations: Technology, economics, and international cooperation. In M. Macauley (Ed.), *Economics and technology in U.S. space policy* (pp. 25–44). (Proceedings of a symposium sponsored by Resources for the Future and the National Academy of Engineering, Washington, DC, June 24–25, 1986.)

McLucas, J. L. (1985, October 7–12). *A multi-national land remote sensing consortium.* Paper presented at 36th International Astronautical Congress, Stockholm, Sweden.

Reinhart, E. E. (1974, May). *Orbit-spectrum sharing between the Fixed-Satellite and Broadcasting-Satellite Services with applications to 12 GHz domestic systems.* (Prepared for the National Aeronautics and Space Administration, R-1463-NASA.) Santa Monica, CA: Rand Corporation.

Rheem, D. L. (1987, March 5). News media push for own "Eyes in the Sky." *The Christian Science Monitor,* p. 1.

Russell, C. S., Harrington, W., & Vaughan, W. J. (1986). *Enforcing pollution control laws.* Washington, DC: Resources for the Future.

chapter 12
Values Underlying Operations Research Models for Synthesizing Communications Satellites

Charles H. Reilly
Associate Professor
Department of Industrial and Systems Engineering
The Ohio State University

INTRODUCTION

The world's first orbiting communications satellite, *Telstar*, was launched by the United States in 1962. Since that time, the number of communications satellites has grown steadily. For example, there were over 200 satellites proposed or deployed in the geostationary orbit alone by the end of 1985 (Jansky & Jeruchim, 1987, p. 7). This proliferation of communications satellites has led to international regulation of satellite communications and to research related to the regulation of satellite communications in such fields as economics, engineering, and operations research. See, for example, Macauley (1984, 1986), Ito, Mizuno, and Muratani (1979), Levis, Wang, Yamamura, Reilly, and Gonsalvez (1988), Reilly (1988), and Reilly et al. (1986a, 1986b, 1988). The subject of this chapter is the operations research (OR) models that have been developed for satellite system synthesis problems, that is, the problems of allocating the orbit/spectrum resource to the satellites and/or satellite administrations that compete for this resource.

Attention will be confined here to the synthesis of communications satellite systems in the geostationary orbit (GSO). This unique orbit is located some 22,000 miles directly above the equator. When a satellite is deployed in the GSO, the satellite has the same orbital period as the earth. Hence, from any point on the earth, a geostationary satellite appears to be fixed in position.

The GSO is a limited natural resource; that is, there are only 360 degrees of GSO. Some satellites that are to serve geographically removed service areas may occupy the same location in the GSO; however, many pairs of satellites cannot share a common location in the GSO because excessive electromagnetic interference would result. Since the longitudes allotted to satellites have to be coordinated in such a way that interference is controlled, the number of satellites that can be deployed in the GSO is limited. By assigning distinct frequencies to potentially interfering satellites, the angular separation between such satellites needed to keep interference in check can be reduced and the effective capacity of the GSO can be increased. Unfortunately, the frequency spectrum is also a limited resource. Hence, even when longitudes are carefully allotted and frequencies are prudently assigned to satellites, the number of satellites that can use any portion of the orbit/spectrum resource is limited. The problem of allocating the orbit/spectrum to the nations of the world in such a way that the anticipated communications needs are met is a difficult and complicated problem.

Most nations (administrations) would prefer to have the exclusive right to use some portion of the orbit/spectrum resource for their own communications needs without any concern for the degradation of their intended signals due to interference. Hence, cooperation among administrations is essential. The International Telecommunications Union (ITU) facilitates this cooperation by bringing representatives from all ITU members together at periodic World Administrative Radio Conference (WARCs) and Regional Administrative Radio Conferences. Solutions to satellite synthesis problems are arrived at through negotiation at these ITU-sponsored conferences. The resulting solutions are politically acceptable, by virtue of their generation through the negotiation process. However, there are many technical issues and tradeoffs that need to be addressed if a truly practical solution is desired.

Some of the technical issues that must be considered are the levels of acceptable single-entry (pairwise) and aggregate interference, the assumed characteristics of transmitting and receiving antennas, the number of channels (frequencies) assigned to each satellite, and the minimum angle of elevation for every administration's satellite. Because several satellite system synthesis problems, with a variety of technical assumptions, may need to be solved before certain tradeoffs (e.g., aggregate interference level versus number of channels per spacecraft) can be evaluated and an acceptable solution can be found, it is useful to have automated procedures that can generate a synthesis solution under a wide range of technical assumptions. To achieve this end, many researchers have recommended OR models, and sometimes solution techniques, for solving satellite system synthesis problems.

One of the difficulties that arises in the application of OR models to real-world problems is the representation of qualitative matters that are not easily expressible in mathematical terms. This difficulty does not diminish the usefulness of the OR models for satellite system synthesis, or for any other practical problem for that matter, but it should lead to negotiations where the solutions to OR models are evaluated on their technical merit and are modified appropriately so that important qualitative issues are adequately addressed.

In this chapter, some of the OR models that have been suggested for satellite system synthesis problems are surveyed. These models will include those that assign channels to satellites, those that allot points on the GSO to satellites, and those that allot segments of the GSO to administrations. Particular emphasis will be placed on the models that seem to be particularly useful for the synthesis of satellites in the Fixed Satellite Service (FSS), since the synthesis of FSS satellites was a major topic at the most recent WARC, WARC-ORB-88. Two examples of OR models for satellite synthesis will be presented in detail. Some of the values inherent in these synthesis models, as well as some of the tradeoffs and conflicts associated with these values, will be discussed. The paper concludes by suggesting how these models might be used at administrative conferences like WARC-ORB-88 as part of an integrated decision-making process that addresses technical issues, as well as the political, social, and economic issues that are not readily representable in OR models.

SURVEY OF OR MODELS FOR SATELLITE SYSTEM SYNTHESIS

Many researchers have studied satellite system synthesis problems and suggested models and/or methodologies for solving them. Before we begin a discussion of some of the OR models that have been recommended for synthesis problems, we need to define what is meant by an OR model and what a satellite system synthesis problem is. An OR model is a mathematical statement of a problem in which a solution is sought that specifies values for decision variables (unknowns), satisfies certain restrictions (constraints), and optimizes some measure of solution merit (objective function). We define a satellite system synthesis problem to be the problem of prescribing some combination of geostationary orbital locations, transmission frequencies, and antenna polarizations for a set of communications satellites that meets stated technical requirements that might include angles of elevation and intersystem interference limits. Not all of the OR models mentioned below are designed to find orbital positions, frequencies, *and* polarizations, but we still consider these to be satellite system synthesis models. Instead of solving for all of these unknowns, certain assumptions about their values are made a priori. For example, models that are designed to find orbital locations for satellites may assume that all satellites use the same frequency(ies) and polarization. This type of assumption is conservative and leads to a solution for a worst-case combination of frequencies and polarizations.

There have been models developed for assigning only frequencies to satellites. Though some of these models may not have been intended for application to satellite systems originally, they may still be useful in a satellite communications setting. These frequency assignment models tend to fall into two classes, depending upon their objective function. Some models are intended to conserve the frequency spectrum. Examples of this type of model include the models of Cameron (1975, 1977), Zoellner and Beall (1977), Baybars (1982), and Levis et al. (1983b). Essentially, these models differ in the number of channels that are assigned to each satellite and in the types of interference restrictions they enforce. Other models, like those of Mathur, Salkin, Nishimura, and Morito (1985) and Allen, Helagson, and Kennington (1987), have objective functions that seek to minimize interference.

There are also models that allot only locations to satellites, normally under the assumption that a common, co-polarized channel is used by all satellites. (For convenience, this type of model will be referred to as a point allotment model.) Ito et al. (1979) have developed a popular nonlinear programming model that attempts to conserve the GSO by positioning satellites in the shortest possible portion of the GSO while still satisfying stated single-entry and aggregate interference requirements. The current version of the software developed for this model is called ORBIT-II. Mizuno, Ito, and Muratani (1984) and Muratani, Ito, and Kobayashi (1988) describe the solution of some synthesis examples using ORBIT-II. Technical details of the approaches used to solve the model of Ito et al. (1979) can be found in manuals from the Kokusai Denshin Denwa Co., Ltd. (Kokusai, 1985, 1988).

Mixed-integer programming models for point allotment problems can be found in Reilly et al. (1986b), Bhasin (1987), Mount-Campbell, Reilly, and Gonsalvez (1986), and Levis et al. (1988). All of these integer programming models utilize minimum required angular separations for all pairs of satellites that can be computed using a procedure developed by Wang (1986). The models use a variety of objective functions that can be included in linear integer programming models for satellite system synthesis. We will be most interested in the objective of minimizing the sum of the absolute deviations in longitude between the locations prescribed for a set of communications satellites and the desired locations of the same satellites. The concept of a desired location is not precisely defined. Each administration could specify a preference for a longitude in the GSO that could serve as the desired location for the administration's satellite, or the desired location of each satellite could be assumed to be the midpoint of its service arc, the set of contiguous longitudes from which the entirety of its intended service area is visible. The desired location concept seems to be very much related to the site value of GSO locations studied by Macauley (1984). This objective function has received attention from all of the authors cited above, and will be included in an example point allotment model presented in the next section.

A promising methodology that has been developed to solve synthesis problems with this objective function when they are modeled as recommended by Mount-Campbell et al. (1986) is described by Gonsalvez (1986) and by Reilly et al. (1988). The present version of the software developed to solve such problems is called OSU-STARS (Orbit/Spectrum Utilization—Synthesis Technique for Allotting Resources to Satellites).

A nonlinear programming model for the problem of allotting locations and assigning frequencies to satellites was suggested by Levis et al. (1983a). Example problems were solved using an extended gradient search procedure by Martin, Gonsalvez, Levis, and Wang (1985). Reilly et al. (1986a) conducted an experiment to assess the performance of two search algorithms on small example problems. They concluded that a cyclic coordinate search seemed to outperform the extended gradient search algorithm in terms of solution quality, as measured by worst-case aggregate interference, but that neither of these methods would be practical on large problems.

Chouinard and Vachon (1981) describe an implicit enumeration method for making frequency and polarization assignments, given fixed orbital positions for the satellites. Orbital positions are also assumed to be fixed in an algorithm by Bove and Tomati (1976) that finds channel and polarization assignments for satellites. Sauvet-Goichon (1976) has developed a procedure that assigns frequencies, then orbital positions, and then polarizations to satellites. After making frequency assignments, a method by O'Leary (1977) establishes orbital positions and antenna polarizations for a set of satellites. Nedzela and Sidney (1985) describe four heuristic algorithms for determining feasible orbital locations, frequencies, and polarizations simultaneously. An interactive synthesis routine that makes channel and polarization assignments and orbital allotments is presented by Christensen (1981). Ottey, Sullivan, and Zusman (1986) suggest several synthesis models with objective functions related to different interference measures that could be solved to find orbital locations, frequencies, and polarizations.

Whyte, Heyward, Ponchak, Spence, and Zuzek (1988) present an algorithm, called NASARC, that can serve as a preprocessor for a synthesis procedure. NASARC finds groups of compatible satellites that can occupy each integral longitude and then assigns groups of compatible satellites to segments of the GSO called predetermined arcs, thereby reducing the burden placed on the normally computationally intensive synthesis routines. None of the satellites are allotted specific orbital positions by NASARC, but much of the work required to find a complete satellite ordering is accomplished. (The term predetermined arc comes from WARC-ORB-85, where it was agreed that all ITU members shall have at least one allotment that includes an orbital position in a predetermined arc, ITU, 1985.)

Kiebler (1985) has suggested the allotment of segments of the GSO to satellite administrations as an alternative to rigid point allotments. Reilly (1988) has formulated an integer programming model for the arc allotment problem and has shown that this problem is related to the point allotment problem addressed by Ito et al. (1979). Kiebler (1985) and Reilly (1988) point out that there are certain advantages that may make the allotment of arc segments preferable to allotments of points on the GSO. In the discussion to follow about the values underlying OR models for satellite system synthesis, both point allotment models and arc allotment models will be considered.

EXAMPLES OF SATELLITE SYSTEM SYNTHESIS MODELS

In this section, we present two OR models for satellite system synthesis for purposes of illustration. One of these example models is for a point allotment problem, and the other is for an arc allotment problem. Both models are mixed-integer programs that would represent a substantial computational challenge for as few as, say, 30 satellites, if they were to be solved for a true (global) optimal solution. In any case, they do represent actual OR models for satellite system synthesis.

A Point Allotment Model

We define the following parameters and decision variables for a mixed-integer point allotment model with the objective function of minimizing the sum of the absolute differences in longitude between the satellites' prescribed and desired locations:

Parameters:

E_j = the easternmost feasible longitude for satellite j.
W_j = the westernmost feasible longitude for satellite j, where it is assumed that $W_j > E_j$.
D_j = the desired location for satellite j.
Δ_{ij} = the minimum required angular separation between satellites i and j that guarantees that a specified single-entry interference requirement is met.

Decision Variables:

r_j = the longitude allotted to satellite j.

$$r_j^+ = \begin{cases} r_j - D_j, & \text{if } r_j > D_j; \\ 0, & \text{otherwise.} \end{cases}$$

$$r_j^- = \begin{cases} D_j - r_j, & \text{if } r_j < D_j; \\ 0, & \text{otherwise.} \end{cases}$$

$$x_{ij} = \begin{cases} 1 & \text{if satellite i is located west of satellite j;} \\ 0 & \text{otherwise.} \end{cases}$$

The point allotment satellite system synthesis model can be formulated as a mixed-integer program as follows:
Minimize

$$z = \sum_j (r_j^+ + r_j^-) \tag{1}$$

Subject to

$$r_j - r_j^+ + r_j^- = D_j \quad \text{for all } j \tag{2}$$
$$r_i - r_j + (E_i - W_j - \Delta_{ij})x_{ij} \geq E_i - W_j \quad \text{for all } i,j \text{ such that } i < j \tag{3}$$
$$-r_i + r_j - (E_j - W_i - \Delta_{ij})x_{ij} \geq \Delta_{ij} \quad \text{for all } i,j \text{ such that } i < j \tag{4}$$
$$r_j \geq E_j \quad \text{for all } j \tag{5}$$
$$r_j \leq W_j \quad \text{for all } j \tag{6}$$
$$r_j, r_j^+, r_j^- \geq 0 \quad \text{for all } j \tag{7}$$
$$x_{ij} \in \{0,1\} \quad \text{for all } i,j \text{ such that } i < j \tag{8}$$

The objective function (1) totals the differences in longitude between the satellites' prescribed locations and their desired locations that are measured in the constraints of type (2). Constraint set (3) guarantees that satellites i and j are sufficiently separated to prevent excessive interference when satellite i is located west of satellite j (i.e., when $x_{ij} = 1$). When satellite j is located west of satellite i (i.e., when $x_{ij} = 0$), constraint set (4) enforces the minimum required separation between satellites i and j. Constraint sets (5) and (6) guarantee that the longitude allotted to each satellite is contained in the satellite's service arc. Finally, nonnegativity and integrality restrictions are enforced by constraint sets (7) and (8).

This mixed-integer programming model is very similar to models that have appeared in Reilly et al. (1986b) and Levis et al. (1988). For a satellite system synthesis problem with n satellites, this model would have $3n$ continuous variables, $n(n - 1)/2$ binary variables, and $n^2 + 2n$ structural constraints. There would be 600 continuous variables, 19,900 binary variables, and 40,400 structural constraints in a 200-satellite problem. Integer programming problems of this size that possess no easily exploited mathematical structure are very difficult to solve for an exact solution, but they frequently can be solved for an approximately optimal solution with a heuristic algorithm.

An Arc Allotment Model

A mixed-integer programming model for the arc allotment problem with the objective function of maximizing the length of the segments of the GSO allotted to each administration has been formulated by Reilly (1988). The following parameters and decision variables are used to construct the model:

Parameters:

E_i = the easternmost feasible longitude for satellite serving administration i.
W_i = the westernmost feasible longitude for satellite serving administration i, where it is assumed that $W_i > E_i$.
F_i = the weighting factor used to determine the length of the segment of the GSO allotted to administration i.

Δ_{ij} = the minimum required angular separation between satellites serving administrations i and j that guarantees that a specified single-entry interference requirement is met.

Decision Variables:

e_i = the eastern edge of the segment of the GSO allotted to administration i.

w_i = the western edge of the segment of the GSO allotted to administration i.

$$y_{ij} = \begin{cases} 1 & \text{if } w_i > w_j; \\ 0 & \text{if } w_i < w_j. \end{cases}$$

(Note that the y_{ij} are only defined for those pairs of administrations for which $\Delta_{ij} > 0$.)

The mixed-integer programming formulation for the arc allotment problem is:

Maximize

$$z = a \tag{9}$$

Subject to

$$w_i - e_i - F_i a = 0 \quad \text{for all } i \tag{10}$$

$$e_i - w_j + (E_i - W_j - \Delta_{ij})y_{ij} \geq E_i - W_j \quad \text{for all } i,j \text{ such that } i < j \text{ and } \Delta_{ij} > 0 \tag{11}$$

$$e_j - w_i - (E_j - W_i - \Delta_{ij})y_{ij} \geq \Delta_{ij} \quad \text{for all } i,j \text{ such that } i < j \text{ and } \Delta_{ij} > 0 \tag{12}$$

$$e_i \geq E_i \quad \text{for all } i \tag{13}$$

$$w_i \leq W_i \quad \text{for all } i \tag{14}$$

$$a \geq 0 \tag{15}$$

$$y_{ij} \in \{0,1\} \quad \text{for all } i,j \text{ such atht } i < j \text{ and } \Delta_{ij} > 0 \tag{16}$$

The objective function (9) maximizes the length of the unweighted GSO arc segment allotted to every administration. Each administration is allotted a weighted-length segment of the GSO based on the length of the unweighted arc segment, a, by constraint set (10). These constraints also ensure that the western endpoint of each arc segment is located west of the arc segment's eastern endpoint. Constraint sets (11) and (12) guarantee that the arc segments allotted to administrations whose satellites might interfere with each other are sufficiently separated. Every location in each allotted arc segment is guaranteed to be feasible for the associated administration by constraint sets (13) and (14). The remaining constraints, (15) and (16), enforce nonnegativity and integrality restrictions on decision variables.

For an arc allotment problem with m administrations, this model includes $2m+1$ continuous variables, and at most, $m^2 + m$ structural constraints and $m(m - 1)/2$ binary variables, with the precise numbers of structural constraints and binary variables depending on the number of pairs of potential interfering satellites (i.e., the number of non-zero Δ_{ij}s). For an arc allotment problem with

150 administrations, there would be 301 continuous variables, and at most, 22,800 structural constraints and 11,175 binary variables. As we saw in the case of the point allotment problem, this arc allotment model cannot be solved easily for a global optimal solution.

These two examples suffice to illustrate the nature of mixed-integer programming models for point allotments. Given alternative objective function selections, a variety of additional models quite similar to those presented above, in terms of their appearance and their dimensions, could be formulated.

VALUES IN OR MODELS FOR SATELLITE SYSTEM SYNTHESIS

There are almost certain to be many opinions about what values are represented in the OR models for satellite system synthesis. The values that are identified here are those perceived by the author. They do not necessarily constitute an exhaustive list of the values underlying OR models for satellite synthesis, nor are they necessarily consistent with the values that might be identified by other researchers who have studied the use of OR models for satellite system synthesis problems. However, some of these values are consistent with the WARC-85 planning principles found in ITU (1985) and Miller (1986). It should be pointed out that these values are not mutually independent, but are interrelated in some cases. Satellite synthesis is further complicated by the fact that some of the these values conflict, essentially giving the problem of synthesizing satellites some of the features of a multiple-criteria optimization problem.

The values inherent in the OR models applicable to FSS satellite system synthesis that have been developed by Ito et al. (1979), Reilly et al. (1986b), and Mount-Campbell et al. (1986) include the following values:

Equitableness. In each of the satellite system synthesis models applicable to FSS satellites, every satellite administration's needs are assumed to be equally important. Since each satellite administration is treated equally, at least in some sense, we say that equitableness is a value that is incorporated into the OR models for satellite system synthesis. A point allotment solution is only feasible if every administration is allotted a longitude in its service arc and has a signal that is sufficiently protected from all potential interferers. Arc allotment solutions are feasible only if every administration is allotted a segment of the GSO that provides them with the same communications capacity per weighting unit (F_i) that all other administrations receive. In either case, every satellite is assumed to be able to transmit signals on all available frequencies. Any synthesis solution satisfying the communications needs of fewer than all ITU members is quite unacceptable.

Equitableness is an important value associated with satellite synthesis models because, if equitableness was not valued, there would be no need to plan the use of the orbit/spectrum resource. Instead, administrations could launch satellites as frequently as they could afford to, could deploy them at the longitudes of their choice, and could effectively prohibit some other administrations from using the orbit/spectrum resource. If equitableness was not an issue, then there might be no ITU or WARCs.

The term equitableness does not imply that the synthesis solutions found using an OR model will be equally pleasing in the view of every administration. There will be some administration that receives stronger interfering signals than all of the other administrations. Also, some administration, perhaps a polar administration, will have the lowest angle of elevation. However, every administration's signal will be adequately protected from interference to the extent provided for in stated technical requirements. Equitableness then implies that no single administration competes for shares of the orbit/spectrum resource at a disadvantage that is attributable to the OR models themselves. Polar nations are dealt a significant disadvantage by nature due to their geography; they simply must have a relatively low elevation angle for their satellites.

Conservation of the GSO. Because the GSO is a limited natural resource, there is some interest in conserving it for future and unanticipated needs, if possible. Some of the measures recommended for evaluating point allotment synthesis solutions suggest that conservation of the GSO is an important issue. For example, in the model formulated by Ito et al. (1979), the objective function is the minimization of the proportion of the GSO occupied by the satellites to which points in the GSO are allotted. Reilly et al. (1986b) and Mount-Campbell et al. (1986) minimize the sum of the distances between prescribed satellite locations and assumed desired locations. Over large land masses, both of these objective functions will tend to force satellites to occupy locations near one another, thereby leaving large gaps over the oceans that would be available for future satellite applications. It might be advantageous though to disperse the satellites as much as possible, potentially leaving some voids in the GSO over land masses. Bhasin (1987) has examined models that favor solutions in which the satellites are more spread out.

By its very nature, the arc allotment model does not conserve the GSO. Rather its intention is to allot as much of the GSO as possible to the satellite administrations. It is then up to each administration to use its allotted arc segment wisely.

Efficient Use of the GSO. As is true for any limited resource like the GSO, there is a risk that the resource will not be used wisely or efficiently. Since the GSO is not an expendable resource, the efficient use of the GSO would involve allocating the resource, in accordance with their communications needs, to administrations that will deploy communications satellites, not simply aspire to do so at some unspecified future date. Therefore, planning the use of the GSO by allotting locations in the GSO to all ITU members may lead to less than efficient use of the precious GSO if all ITU members do not use their allotments because either (a) they never launch a satellite, or (b) they have no real need for their own satellite(s).

In the case of point allotments, the objective functions suggested by Ito et al.

(1979), and by Reilly et al. (1986b) and Mount-Campbell et al. (1986), would tend to mitigate the inefficiency in the use of the GSO, because adjacent satellites serving service areas on the same land mass tend to be separated just far enough to protect their signals from interference. Since the satellites are usually not spread out, leaving most administrations with reasonably high angles of elevation, the cost of inefficiency, as measured by signal quality and system cost, is not as substantial as it might be.

With arc allotments, the penalty for inefficiency is more severe because the benefits of arc allotments, primarily operational flexibility and the ability to deploy many satellites, are diminished if all allotments are not fully utilized.

Flexibility. One value issue that is considerably better addressed with arc allotments, as opposed to point allotments, is the issue of flexibility. By allotting each satellite administration an arc segment instead of a single point on the GSO, the administration can then select the best longitude in its arc segment for its satellite, using any criterion it deems appropriate. It can deploy as many satellites as can be adequately accommodated in its arc segment. Furthermore, satellites can be repositioned as the communications needs of the administration change.

Point allotment solutions are quite inflexible. Specific longitudes are reserved for particular satellites. Any movement of a single satellite could result in excessive interference being received in a number of service areas that are served by nearby satellites.

Computational Practicality. The problem of allotting locations or GSO arc segments to satellites is a very difficult problem to solve. In face, this problem is probably *NP-Hard,* that is, among the most difficult optimization problems one could face (Garey & Johnson, 1979). It is doubtful that the best (that is, the best in a mathematical sense) synthesis solution can be found via any solution technique in a reasonable amount of time. As a result, methodologies that produce heuristic, or approximate, solutions may represent the most practical choices for solving satellite system synthesis models.

Some of the OR models for satellite synthesis problems have been formulated in such a way that their structure can be exploited in a solution procedure so that some solution, hopefully a good solution, can be found in a reasonable amount of time. See, for example, the formulation by Mount-Campbell et al. (1986), whose structure is exploited in the solution procedure suggested by Gonsalvez (1986). A sophisticated synthesis model that is so general as to make solving it a practical impossibility is of little use. In order to find workable synthesis solutions, models that are, at a minimum, solvable by heuristic means for approximate, but feasible, solutions are a must.

VALUE TRADEOFFS AND VALUE CONFLICTS

Given the values listed in the last section, it is no wonder that there are value tradeoffs and conflicts that need to be considered before an acceptable synthesis solution is found. In this section, we will discuss a few of the tradeoffs and conflicts associated with the values enunciated earlier. This discussion only illustrates the tradeoffs and conflicts, and in no way does it characterize all of the value tradeoffs and conflicts that might arise when synthesizing communications satellites.

Some compromise between equitableness and efficient use of the GSO must be reached if a good synthesis solution is to be found. Though it is equitable to include all ITU members in the planning of allotments for satellite communications, it is certainly not efficient to designate certain locations in the GSO, some of them perhaps prime locations over large land masses comprised of many ITU-member nations, to *belong* to administrations that may never deploy a single communications satellite of its own.

Hudson (1986) suggests a sort of statute of limitations for deploying satellites. She recommends that a time limit be imposed so that any allotments not used in a reasonable amount of time are returned to a pool so they can be re-allotted to other administrations. Jansky (1988) also warns of the abuse of the GSO resulting from planning and rigid regulation of satellite communications.

The value of conservation of the GSO and flexibility are in conflict with one another as well. Conserving portions of the GSO for new and unforeseen communications needs limits the number of potentially feasible synthesis solutions, some of which might provide administrations with more opportunities to adapt their satellite systems to their own needs as they evolve over time. Conservation also limits the size of the arc segments that might be allotted to ITU members. However, according to the ITU (1985) the present and anticipated needs of the ITU members will be considered at WARC-88 before any allowance is made for unforeseen requirements.

Finally, computational practicality may come at the expense of efficient use of the GSO. If single-entry interference restrictions are enforced in place of or in addition to aggregate interference constraints, the prescribed angular separations between adjacent satellites can only be increased. As a result, fewer satellites can be accommodated in the GSO. The enforcement of aggregate interference requirements leads to nonlinear functions in OR models for satellite synthesis (Ito et al., 1979; Levis et al., 1983a). Optimization in the presence of nonlinear functions is typically more difficult than it is in the presence of linear functions only, due to numerical stability problems.

ORBIT-II attacks a synthesis problem in which aggregate and single-entry interference constraints are enforced. Though ORBIT-II does solve a nonlinear optimization problem, it does so by solving a series of linear programming problems that provide successively better approximations to the nonlinear problem of interest (Kokusai, 1984, 1988). Furthermore, ORBIT-II prohibits the reordering of the satellites to avoid some mathematical singularities. The method suggested by Gonsalvez (1986) that is used in OSU-STARS enforces only single-entry interference constraints, but it considers many more orderings of the satellites than ORBIT-II does.

Certainly additional value tradeoffs and conflicts could be listed here, as could some technical tradeoffs. Those already mentioned underscore the difficulty of satellite synthesis problems and the importance of the OR models that facilitate the evaluation of the technical merit of candidate synthesis solutions in a reasonable time.

CONCLUSION

The problem of synthesizing 200 to 300 communications satellites is an important problem that must be dealt with by all nations in unison. The problem is too large and far too complicated for the ITU to rely on manual solution strategies for arriving at an acceptable synthesis solution. Furthermore, since many synthesis scenarios, with different technical assumptions, may be needed to be considered, many solutions may have to be found in a reasonable amount of time. Naturally, this suggests automated methods for solving satellite system synthesis problems that can be exercised each time a new set of technical parameters is to be considered. The OR models that have been suggested for satellite system synthesis, especially those computerized in ORBIT-II and OSU-STARS, provide means for finding solutions to synthesis problems economically.

One potential shortcoming of the OR models that have been suggested for satellite system synthesis is that some of the nonquantitative issues, especially political, social, and economic issues, are not easily expressible in mathematical terms, as they would have to be in order to be addressed directly by the OR models. Since solutions to satellite system synthesis problems can be found reasonably economically, characteristics of one solution that are found to be undesirable may be removed by solving another synthesis problem with slightly modified technical requirements. For example, if one satellite serving a developing nation would be allotted an orbital position that has such a low angle of elevation that extraordinary investments in equipment would be necessary to make this position useful, another synthesis problem in which this satellite is assumed to have a shorter service arc could be solved to determine whether this undesirable situation can be easily rectified without detrimentally affecting the service provided the other administrations.

The OR models for satellite system synthesis are not a substitute for the negotiators at WARCs. Rather, they represent useful tools that complement the negotiation process by quickly determining the technical merit of the solutions found under different sets of technical assumptions. Together with the negotiators, the OR models for satellite system synthesis provide an integrated decision-making structure that recognizes the importance of both technical and qualitative issues. As a consequence of the application of OR models, the merits of new synthesis scenarios can be accurately assessed without speculation, and the prospects for reaching an acceptable synthesis solution are ameliorated.

REFERENCES

Allen, J., Helagson, R., & Kennington, J. (1987). The frequency assignment problem: A solution via nonlinear programming. *Naval Research Logistics, 34* (1), 133–139.

Baybars, I. (1982). Optimal assignment of broadcasting frequencies. *European Journal of Operational Research, 9* (3), 257–265.

Bhasin, P. (1987). *Mathematical programming formulations for satellite synthesis.* Unpublished master's thesis, The Ohio State University.

Bove, F., & Tomati, L. (1976). A planning method for television broadcasting from satellites. *European Broadcasting Union Review,* Number 158.

Cameron, S. (1975). Sequential insertion: An algorithm for conserving spectrum in the assignment of operating frequencies to electronic systems (ECAC-TN-75-0023). Unpublished manuscript.

Cameron, S. (1977). The solution of the graph-coloring problem as a covering problem. *IEEE Transactions on Electromagnetic Compatibility,* EMC-19(3), 320–322.

Chouinard, G., & Vachon, M. (1981). *A synthesis of a plan by computer using an assignment method and inductive selection.* ITU/CITEL/DOC-Canada Preparatory Seminar for 1983 RARC, Ottawa, Canada.

Christensen, J. (1981). *BSS CAPS: a system description.* ITU/CITEL/DOC-Canada Preparatory Seminar for 1983 RARC, Ottawa, Canada.

Garey, M., & Johnson, D. (1979). *Computers and intractability: A guide to the theory of NP-completeness.* New York: W. H. Freeman and Co.

Gonsalvez, D. (1986). *On orbital allotments for geostationary satellites.* Unpublished doctoral dissertation, The Ohio State University, Columbus, OH.

Hudson, H. (1986). Access to information resources: The developmental context of the space WARC. In D. Demac (Ed.), *Tracing new orbits: Cooperation and competition in global satellite development.* New York: Columbia University Press.

International Telecommunications Union. (1985). *Report to the Second Session of the Conference. Addendum.* World Administrative Radio Conference on the Use of the Geostationary Satellite Orbit and the Planning of Space Services Using It, First Session, Geneva.

Ito, Y., Mizuno, T., & Muratani, T. (1979). Effective utilization of geostationary orbit through optimization. *IEEE Transactions on Communications,* COM-27 (10), 1551–1558.

Jansky, D. (1988). *The use and abuse of the geostationary orbit* (pp. 19–23). AIAA 12th International Communication Satellite Systems Conference, Arlington, VA.

Jansky, D., & Jeruchim, M. (1987). *Communication satellites in the geostationary orbit* (2nd ed.). Boston: Artech House.

Kiebler, J. (1985). *NASA studies on the U.S. approach for fixed satellite service management at the 1985/88 Space WARC.* (Document AH-178-299). Ad Hoc Committee 1978 of the Interdepartmental Radio Advisory Committee, Washington, DC.

Kokusai Denshin Denwa Co., Ltd. (1984). *The Orbit Spacing Minimizer (ORBIT-II) user's manual.* Tokyo: Author.

Kokusai Denshin Denwa Co., Ltd. (1988). *ORBIT-II program (Version 3.87): theoretical manual.* Tokyo: Author.

Levis, C., Martin, C., Gonsalvez, D., & Wang, C. (1983a). *Engineering calculations for communications satellite systems planning.* (Second Interim Report 713533-3 for

Grant NAG 3-159). Columbus, OH: The Ohio State University, ElectroScience Laboratory.

Levis, C., Martin, C., Wang, C., & Gonsalvez, D. (1983b). *Engineering Calculations for Communications Satellite Systems Planning.* (First Summary Report 713533-2 for Grant NAG 3-159). Columbus, OH: The Ohio State University, ElectroScience Laboratory.

Levis, C., Wang, C., Yamamura, Y., Reilly, C., & Gonsalvez, D. (1988, August). The role of service areas in the optimization of FSS orbital and frequency assignments. *IEEE Transactions on Electromagnetic Compatibility,* EMC 30(3), 371–379.

Macauley, M. (1984). *The site value of locations in the geostationary orbit.* (Resources for the Future, Discussion Paper EM85-01). Washington, DC: Resources for the Future.

Macauley, M. (1986). The contribution of a partnership between economics and technology. In M. Macauley (Ed.), *Economics and technology in U.S. space policy.* Washington, DC: Resources for the Future.

Martin, C., Gonsalvez, D., Levis, C., & Wang, C. (1985). Engineering calculations for communications satellite systems planning. (Interim Report 413533-4 for Grant NAG 3-159.) Columbus, OH: The Ohio State University, ElectroScience Laboratory.

Mathur, K., Salkin, H., Nishimura, K., & Morito, S. (1985). Applications of integer programming in radio frequency management. *Management Science, 31* (7), 829–839.

Miller, E. (1986, March). Impact of the 1985 Space World Administrative Radio Conference on Frequency/Orbit Planning and Use. *AIAA 11th Communications Satellite Systems Conference,* San Diego, CA.

Mizuno, T., Ito, Y., & Muratani, T. (1984). Computer tools for optimizing orbit use. *AIAA 10th Communications Satellite Systems Conference* (pp. 549–557). Orlando, FL.

Mount-Campbell, C., Reilly, C., & Gonsalvez, D. (1986). *A mixed integer linear programming formulation of the FSS synthesis problem using minimum required pair-wise separations* (Working Paper 1986-006). Columbus, OH: The Ohio State University, Department of Industrial and Systems Engineering.

Muratani, T., Ito, Y., & Kobayashi, H. (1988). Study of allotment planning for equitable and efficient use of geostationary orbit. *AIAA 12th International Communication Satellite Systems Conference* (pp. 24–31). Arlington, VA.

Nedzela, M., & Sidney, J. (1985). Méthodes de planification pour la radiodiffusion par satellites. *INFOR, 23* (2), 100–119.

O'Leary, T. (1977). Satellite broadcasting network—A fast planning method. *European Broadcasting Union Review,* Number 161.

Ottey, H., Sullivan, T., & Zusman, F. (1986). Optimization techniques applied to spectrum management for communications satellites. *AIAA 11th Communication Satellite Systems Conference* (pp. 199–210). San Diego, CA.

Reilly, C. (1988, July). A satellite system synthesis model for orbital arc allotment optimization. *IEEE Transactions on Communications,* COM 36(7), 845–849.

Reilly, C., Mount-Campbell, C., Gonsalvez, D., Martin, C., Levis, C., & Wang, C. (1986a). Broadcasting satellite synthesis using gradient and cyclic coordinate search procedures. *AIAA 11th Communication Satellite Systems Conference* (pp. 237–245). San Diego, CA.

Reilly, C., Mount-Campbell, C., Gonsalvez, D., & Levis, C. (1986b). *Alternative mathematical programming formulations for FSS synthesis* (Working Paper 1986-005). Columbus, OH: The Ohio State University, Department of Industrial and Systems Engineering.

Reilly, C., Gonsalvez, D., & Mount-Campbell, C. (1988). A k-permutation algorithm for fixed satellite service orbital allotments, *AIAA 12th International Communication Satellite Systems Conference* (pp. 46–52). Arlington, VA.

Sauvet-Goichon, D. (1976). A method of planning a satellite-broadcasting system. *European Broadcasting Union Review,* Number 155.

Wang, C. (1986). *Optimization of orbital assignment and specification of service areas in satellite communications.* Unpublished doctoral dissertation, The Ohio State University, Columbus, OH.

Whyte, W., Heyward, A., Ponchak, D., Spence, R., & Zuzek, J. (1988). Numerical arc segmentation algorithm for a radio conference—A software tool for communication satellite systems planning. *AIAA 12th International Communication Satellite Systems Conference* (pp. 32–45). Arlington, VA.

Zoellner, J., & Beall, C. (1977). A breakthrough in spectrum conserving frequency assignment technology. *IEEE Transactions on Electromagnetic Compatibility, EMC-19* (3), 313–319.

IV
Economic Development and International Comparisons

chapter 13
Telecommunications Technologies and Social Change: The Japanese Experience

Tetsunori Koizumi
Associate Professor of Economics
Department of Economics
The Ohio State University

INTRODUCTION

People on the ground, dressed in formal attire, are looking up towards the sky from the terrace of a mansion overlooking what appears like the San Francisco Bay. The Stars and Stripes, on the flagpole of the mansion, is waving gently in the breeze. Another group of people, also dressed in formal attire, are saluting (one gentleman with his hat off) to the onlookers from the balloons floating in the air. On the right is a white rectangle that contains an inscription, written in black, which reads "America."

Never mind that the picture contains inaccuracies: The Stars and Stripes, for example, has only 12 stars and 11 stripes. Never mind that the picture is a bit incongruent: Despite what appears to be a festive occasion, judging from the attire of the people on the terrace and on the balloons, the rest of the town looks completely deserted and is enveloped in eerie silence. After all, the artist, Yoshitora, never set foot on American soil; depicting "life in America" was not what Yoshitora intended to do with this picture. Rather, Yoshitora intended this picture, we are told, to be an expression of his amazement at the fact that balloons, carrying people in the air, were being used in America as a means of communication over a long distance.

From the beginning of Japan's contact with the West, the Japanese have showed their fascination with Western communications technologies—with the very important exception of the Tokugawa Shogunate. When Commodore Perry, during his second trip to Japan which resulted in the signing of the Kanagawa Treaty in 1854, presented a telegraph machine to the Shogunate, Tokugawa officials, even after they had been treated to an amazing demonstration of what

217

the machine invented by the American painter Samuel Morse could do, were unimpressed with this foreign gadget and stored it away in one of the warehouses of the Shogunate. The upstart feudal lords such as Nabeshima of Saga and Shimazu of Satsuma, on the other hand, immediately saw the potential of the machine that "spewed black dots on a white paper" for building the infrastructure of a new nation. This episode thus foreshadowed the fall of the Tokugawa Shogunate and the rise of the new Meiji leaders, who were far more receptive to foreign ideas and technologies.

One of the new Meiji leaders, Hirofumi Ito, demonstrated his receptiveness to foreign technologies in a speech he delivered at the reception given in honor of the visiting Iwakura Mission in San Francisco on January 15, 1872. In this speech Ito unveiled a national plan which his new nation would pursue in order to develop the infrastructure essential for industrialization. The speech contained reference to four concrete technologies (railroad, telegraph, lighthouse, and shipbuilding) which, all related to communications in one way or another, would constitute the kernel of that infrastructure. And the Meiji government, with such innovative leaders as Ito, who served as the first prime minister of a new constitutional government, quickly took to the task of developing these technologies.

The Japanese fascination with communications technologies continues to this day, some hundred years after Yoshitora painted the picture of floating balloons. Needless to say, the Japanese no longer have to "compose" the picture of "America" from traveller's tales. Direct Broadcasting by Satellite (DBS), which officially began on July 4, 1987, now carries vivid pictures of life in America on the television screens across the nation. Thanks to this latest communications technology that floats in the high air, the Japanese now have direct access to life in America.

Advances made in communications technologies in the last hundred years have brought about remarkable changes in the way life is organized in many societies, including Japan. The latest addition to the arsenal of telecommunications technologies such as DBS is on the verge of bringing in even more dramatic changes. And Japan, having grown out of its traditional role as an importer of foreign technologies, is now very much an active participant of social change in the global information society. Thus, examining how the Japanese have dealt, and are dealing, with the problem of social change induced by communications technologies offers an interesting case study in thinking about the general question of telecommunications, values, and the public interest.

THE TYPOLOGY OF EVOLUTIONARY CHANGE

What is the nature of social change that recent advances in telecommunications technologies have triggered, and are inducing, in the global information society?

If we are to derive meaningful answers to this question, we need a frame of reference. Thus, we begin our inquiry with a discussion of what might be termed the *typology of evolutionary change.*

We distinguish among four types of evolutionary change: (a) growth and decay, (b) structural change, (c) synergistic transformation, and (d) transmutation. This typology of evolutionary change is meant to be general and applies to biological as well as societal evolution.

The first type of change, *growth and decay,* has to do with the quantitative change of living systems, which include *social systems* as the term is defined by Miller (1978). For example, a biological species grows or decays as indicated by the size of its population. *Structural changes,* on the other hand, refer to changes in the composition of parts, sectors, or subsystems. The two types of evolutionary change, growth and decay and structural change, can be complementary. For example, we are all familiar with the structural changes in the economy from the agriculture sector to the manufacturing and service sectors that accompanied modern economic growth.

The third type of evolutionary change, *synergistic transformation,* refers to the interaction among parts, sectors, or subsystems that results in a new type of phenomenon at the aggregate level. The existence of a synergistic transformation has long been recognized in the natural sciences, e.g., chemistry, and has been known as the quantity–quality transformation. The Hegelian dialectics offers an example of the idea of a synergistic transformation as applied to the realm of the human sciences. The fourth type of change, *transmutation,* refers to the appearance of a new phenomenon as a result of a sudden change in the environment of a living system. Thus a sudden change in the climate triggers changes in the genotypes which in turn lead to changes in the phenotypes. In the case of societal evolution, a transmutation occurs as a result of one social system interacting with other social systems, or of a sudden burst of a new archetype of the collective unconscious which constitutes an important part of the environment for a social system (Jung, 1969).

All four types of evolutionary change are relevant to social change, and all societies have seen changes of both growth and decay and structural change varieties since the Industrial Revolution. The start of the Industrial Revolution was, of course, a period marked by a synergistic transformation, in that the whole episode of industrialization was triggered by the merging of science and business. But the drive towards standardization and mechanization which characterized what Toffler (1980) calls the Second Wave civilization was mainly responsible for the prevalence of the first two types of evolutionary change.

The development of telecommunications technologies, which is one aspect of the emergent Third Wave, has brought the last two types of evolutionary change into prominence. The merging of telephones and computers, which has ushered in the age of *compunication,* is an outstanding example of synergistic transformation in the realm of technologies which has a corresponding implication for social change. The telephone itself has gone through a series of interesting transmutations. At first, the telephone was viewed as part of the infrastructure of a modern industrial nation, as was indeed the case in Meiji Japan when it was first introduced in 1890. This period was followed by a long period, which extends well after World War II in the case of Japan, in which having a telephone was viewed as a status symbol. This defines the second phase of transmutation, which was soon superseded by the third phase in which the telephone became a necessity of modern life. The telephone is now undergoing the fourth phase of transmutation, from a modern instrument of communication into a strategic technology of value creation in business.

The four types of social change have their counterparts in the realm of human values. Here, of course, we must accept the possibility of causality working in both directions. On the one hand, a value change, initiated by an individual or a group of individuals, may escalate itself into a social change involving a large number of societal members. On the other hand, a social change taking place at the aggregate level in the form of policy change, or a contact with other societies, may induce a change in the life of individuals and their values. The Weberian thesis that the rise of capitalism in the West was triggered by the development of the Protestant ethic offers an example of how a widespread social change can be induced by a value change initiated by a small number of influential people (Weber, 1930). On the other hand, modern economic growth, raising the standard of living of the general public, has spread the middle-class value system in the so-called advanced industrial societies. The structural change in the economy from agriculture to industry also involved a value change in the form of the rise of *city culture,* or the set of values we associate with city life. Thus, value changes are expected to take place whenever there are growth and decay and structural change.

Value changes associated with synergistic transformation and transmutation are naturally far more dramatic than those associated with growth and decay and structural change, for increases in complexity and diversification are usually associated with these types of evolutionary change. In fact, we are in the midst of a social change full of synergistic transformations and transmutations. To see more clearly the linkage between technological development and value change, we need some insight into the workings of a *social system.*

TECHNOLOGY AND VALUE CHANGE: A SOCIAL SYSTEMS PERSPECTIVE

For the last 200 years, since the Industrial Revolution, technology has been a dominant agent of social change in most societies, giving rise to what Ellul (1964) calls the *technological society.* Recent advances in telecommunications technologies have extended the reach of the technological society to include all societies of the world, opening up the new era of cultural confrontations and value conflicts. The question of technology and human values must therefore be discussed in the context of the global information society. For this purpose, we first develop a social systems analysis which can be used to analyze how values are affected by technological development and how different societies come into conflict with one another over the question of values.

Society can be conceived as existing in the space of all human action, following the sociological tradition as represented by the works of Talcott Parsons (Parsons & Shils, 1952). With this conception of society as an action space, society can be regarded as consisting of three subsystems of *culture, economy,* and *polity.* By culture is meant the set of all human actions pertaining to the creation and dissemination of ideas, values, and symbols; by economy, the set of all human actions pertaining to the production and distribution of goods and services; and by polity, the set of all human actions pertaining to the maintenance of the social order and cohesion.

Science, which has to do with the discovery of regularities and laws in nature and society, can be regarded as a subset of culture in our conception of the social system. Technology, on the other hand, traditionally had to do with the sustenance of life and can therefore be regarded as having belonged to economy. In modern technological societies, however, technology is ubiquitous in all realms of social life. Thus, some technologies belong to culture, others to economy, and still others to polity. Knowing what type of technologies belong to what subset of the social system is important in answering the question of what type of social and value change would be induced by what kind of technologies. Needless to say, the same technology can have different impacts on the social system as the society goes through changes, thus giving rise to transmutations. For example, the telephone belongs to polity when it is used as a means of social communication by the government, to culture when it is perceived mainly as a status symbol, and to economy when it is used as a strategic instrument of value creation in business.

From this general discussion of technology and value change, we can now move on to a discussion of the impact of telecommunications technologies on society. Although some people may feel hesitant to characterize what is happening around us as the telecommunications revolution, there are indeed some revolutionary aspects to telecommunications technologies. In particular, these technologies promise to bring about many instances of synergistic transformations and transmutations, if they have not already. The combined effects of all these synergistic transformations and transmutations may well turn out to be revolutionary, just as the Industrial Revolution involved many synergistic transformations and transmutations in the social system.

One thing that is definitely revolutionary about the telecommunications technologies is the speed with which data are gathered, stored, processed, and transmitted. As a result, information has become a strategic good which promises high value-added in the economic arena. In fact, information is transmuting itself from a public into a private good.

The full transmutation of information from a public into a private good is not yet complete. But the fundamental restructuring of the economy is already under way, transforming industrial societies into information societies. One illustration of this restructuring is the way the telephone is used in business. We have already pointed out how the merging of telephones and computers has brought about a transmutation of the telephone. The telephone is no longer just a machine that keeps people in touch with each other; it is a technology that adds to the productivity of business. Thus the telephone bill is not just another item of cost but an expenditure on R&D.

The transmutation of the telephone as technology has transformed the telephone industry from a rather unexciting public monopoly into a highly competitive information industry. Moreover, the stage of competition has expanded to include the whole world. As a result, the industry has become a strategic player in a nation's trade policy and added a new dimension to the public policy debate.

A widespread use of telecommunications technologies is expected to induce significant social and value changes. For one thing, the development of telecommunications technologies has opened up a new channel of interaction between producers and consumers in the marketplace. Helped by flexible, computer-assisted production and a global information network, the producer can now design his or her product very much to the liking of the individual consumer. Thus, the consumer, instead of having to choose from a limited array of standardized products, can now have some say in the kind of product he or she wishes to purchase from the producer. The natural consequence of this expanded scope of the consumer sovereignty would be the diversification of consumer demand and the diffusion of values. Thus, we are looking at the possibility of technological development in economy inducing value changes in culture.

In addition to the diffusion of values induced by the diversification of consumer demand in the marketplace, there has been a tendency towards the diffusion of cultural values, especially in liberal democratic societies. One consequence of such a diffusion of values is the increase of special interest groups in the political arena. These special interest groups today have access to culture in the form of cable televisions which can be used to promote their value system. Political life will therefore be fundamentally altered by the widespread use of telecommunications technologies. Thus, technological development in economy can have a profound impact on polity as well.

TELECOMMUNICATIONS TECHNOLOGIES AND SOCIAL CHANGE: THE JAPANESE EXPERIENCE

What type of social change is induced by what kind of technology depends on a specific pattern of interaction among the subsystems of culture, economy, and polity. That pattern of interaction will differ from one society to another, be-cause the way in which the social system operates is affected by such factors as historical traditions, societal norms and cultural values. The purpose of this section is to derive what might be termed the *Japanese model* of social change from history and to see if the model applies to the recent episode involving telecommunications technologies.

While there is a fair amount of consensus regarding the significance of the Meiji Restoration in the modernization of Japan, there is some debate as to the nature of social change surrounding this historical event. According to one school of historians and social scientists, the Meiji Restoration signifies the emergence of a new type of society, though that new society represents the culmination of an internal social change which had been building up within Tokugawa Japan. In the context of our typology of evolutionary change, the emergence of Meiji Japan is an instance of synergistic transformation, in that a change in the pattern of interaction among social subsystems as represented by the four social classes in Tokugawa Japan gave rise to a new type of social system. There are other historians and social scientists, however, who regard the Meiji Restoration as more of a transmutation triggered by external shocks, more of an ecosystemic reaction to the events taking place in the world from which Japan could no longer be insulated.

From a social systems perspective, these two types of change are not mutually exclusive. In fact, it is best to consider the social change surrounding the Meiji Restoration as involving elements of both synergistic transformation and trans-mutation. An external shock in the form of the arrival of Commodore Perry with his black ships was surely an event that rudely woke up the Japanese to the realities of the outside world, accelerating the synergistic transformation of their society that had already been going on in Tokugawa Japan and forcing trans-mutations of their social institutions in order to catch up to the West. However, for a nation such as Japan, which has a long history of importing and adapting foreign ideas and technologies, a radical social change following an external shock of one sort or another is actually an established pattern of social change and may therefore be called the Japanese model of social change (Koizumi, 1982). The Meiji Restoration and World War II are, of course, two dramatic examples of the Japanese model of social change in modern history.

The social change which the development of telecommunications technolo-gies has induced is no exception to this model and involves elements of both synergistic transformation and transmutation. What triggered the transmutation of Japanese society in this instance was a series of external shocks (the 1971 Nixon shock and the 1973 oil shock, in particular) that exposed the fragile nature of Japan's industrial prowess in the world economy.

The Japanese were quick to translate the realization of their vulnerability to external events into concrete agenda for social reform. As has become the established pattern since the Meiji Restoration, the government took the initiative in the task of transforming Japan to meet the new challenges. The Ministry of International Trade and Industry (MITI) quickly summoned the Industrial Structure Council and charged it with the task of developing a new "vision" for Japan, a new direction in the restructuring of the Japanese economy. The Industrial Structure Council is a rather powerful group of Japanese politics whose influence on policy-making in Japan is sometimes compared to that of the National Security Council in the United States. It was this group, for example, that advised MITI in the early 1960s to establish Japan Electronic Computer Corporation so that Japan could compete effectively in the area of electronics with foreign companies such as IBM and Texas Instruments.

What the Industrial Structure Council came up with in response to the new challenges facing Japan in the turbulent world economy was the vision of the Information Society. The vision called for the restructuring of the economy from the hitherto energy-intensive industries to the information-oriented industries. The advice was quickly followed up by MITI with a new industrial policy. According to one calculation done by the Economic Planning Agency, the amount of *information* produced in the Japanese economy increased from 5.5% of GNP for 1977 to 10% of GNP for 1984, one evidence of the swiftness with which the Japanese took to the task of transforming their society into an information society.

Japan is currently in the second phase of this synergistic transformation (or *informationalization*, as the Japanese call it) of its society, which includes a fundamental restructuring of the telecommunications industry. A single event that signifies this restructuring, a synergistic transformation of the relationship between government and business, is the passing of the Telecommunications Business Law in 1984, which paved the way for the privatization of Nippon Telegraph and Telephone (NTT).

NTT was established as a public monopoly in 1953 because it was felt at the time that the main mission of a telephone company was to make available the uniform communications service to as many Japanese as possible. Information as represented by what the telephone company provided was regarded as a public good. With the merging of telephones and computers, which ushered in the new age of information as a private good, NTT started its second life as a private corporation on April 1, 1985. Its new mission was to transform Japan into a *new total telecommunity* while offering a comprehensive service in data communications to the public as well as the private sector.

NTT is playing an active role in transforming Japan into an *information network society* today. For example, in September 1984, it started an experiment called the Information Network System (INS) in a suburb of Tokyo to test the feasibility of converting Japan into an information network society by the year 2000 with a nation-wide network of digital communications. It is an innovative social experiment but also one that reflects the Japanese concern over the integrity of their traditional values even in the age of high technology. The main purpose of the experiment is to see whether the information network system helps to maintain and, hopefully, enhance the stability of social life, especially in the case of a natural disaster such as an earthquake. It is an interesting social experiment, in that it will test the importance of direct contact, which the Japanese have always cherished as represented by an old saying: "Touching of the sleeves is also a sign that you and I are not strangers," in maintaining the cohesion of social life.

NTT, in cooperation with the Ministry of Posts and Telecommunications, is also cultivating the so-called *Captain System* (Character and Pattern Telephone Access Information Network System). This is the Japanese version of the "videotex" system in other countries. A large-scale experiment in the Captain System also began in 1984 with the database consisting of some 200,000 videos. These videos cover a range from news and weather reports to shopping and town guides, and from legal advice to news about hot springs. This experiment, which basically alters the way people conduct their social life, promises to bring about some fundamental changes in the Japanese value orientations (Caudill, 1962).

CONCLUSION

One consequence of the maturation of Japan as an industrial society was the propagation of the middle-class value system. This is confirmed by government-conducted opinion surveys which show over 90% of the Japanese responding as belonging to the middle class in recent years. Their claim seems to be justified, judging, for example, from the ownership of major consumer durables.

An increase in the standard of living inevitably leads to a value change. One area where this change has occurred is in decreased Japanese interest in savings, and in the new Japanese attitude towards present versus future consumption. The government has taken a major step in encouraging present consumption by abolishing favorable tax treatment of interest income.

Closely related to this is the value change concerning the conception of work. An amendment to the 40-year-old Labor Standards Laws went into effect in September 1987, making the 40-hour work week official and gradually phasing out the 48-hour work week. The short-run impact of this change in the labor practice, as anticipated by the government, is to stimulate consumption of imported goods, which will help lessen the trade friction with Japan's trading partners. But it will turn out to be a seminal event which fundamentally alters the Japanese attitude towards work. In fact, a survey conducted by the Prime Minister's office indicates that, since 1983, spending more time for leisure has become the priority of life for most Japanese.

Whether these value changes which are already under way will lead to a major change in culture will depend on whether these changes, which are mostly limited to changes in economy and polity, reflect, or will lead to, changes in the value system in the deep structure of the Japanese mind. One effect of development of telecommunications technologies is to make it impossible for the Japanese to retreat into the policy of cultural isolation. In fact, no social system is immune from the vivid images of life in foreign lands displayed on television screens, though the Japanese, for now, have soundly rejected *Dallas* and the value system it represents.

The Japanese government will no doubt continue to play out its Confucian commitment to maintain the cohesion of the social system. But, in some areas, there are signs that the effectiveness of the government as a *pattern system* (Kuhn, 1974) is being eroded by events taking place in the global information society. Perhaps Shoin Yoshida, who instilled a rebellious spirit in his pupils, such as Hirofumi Ito, Koin Kido, and Aritomo Yamagata, who grew up to become influential leaders of Meiji Japan, correctly anticipated what modernization was all about when he proclaimed: "Neither the lords nor the Shogun can be depended upon to save the country, and so our only hope lies in grass-roots heroes." As for the global information society we live in, neither the national governments nor the international organizations can be depended upon to save us from the current chaos and confusion of values. Hopefully, a new breed of grass-roots heroes are growing up in our schools.

REFERENCES

Caudill, W. (1962). Japanese value orientations. *Ethology, 1,* 53–91.

Ellul, J. (1964). *The technological society.* New York: Knopf.

Jung, C. G. (1969). *The archetypes of the collective unconscious.* Princeton, NJ: Princeton University Press.

Koizumi, T. (1982). Absorption and adaptation: Japanese inventiveness in technological development. In S. B. Lundstedt & E. W. Colglazier (Eds.), *Managing innovation: Social dimensions of creativity, invention and technology* (pp. 190–206). New York: Pergamon Press.

Kuhn, A. (1974). *The logic of social systems.* San Francisco, CA: Jossey-Bass Publishers.

Miller, J. (1978). *Living systems.* New York: McGraw-Hill.

Parsons, T., & Shils, A. (Eds.). (1952). *Toward a general theory of action.* Cambridge, MA: Harvard University Press.

Toffler, A. (1980). *The third wave.* New York: William Morrow and Company.

Weber, M. (1930). *The Protestant ethic and the spirit of capitalism.* London: Allen and Unwin.

chapter 14
Value Issues in Telecommunications Resource Allocation in the Third World

Rohan Samarajiva
Assistant Professor & Ameritech Faculty
Fellow, Department of Communication
The Ohio State University

Peter Shields
Research Associate, Center for
Advanced Study in Telecommunication
The Ohio State University

INTRODUCTION

The past two decades have witnessed a qualitative change in the perception of the role played by information-communication technologies in society. They are now seen by many as the engines of growth and as the leading sectors of Western economies (e.g., Bell, 1973; Porat, 1977). The disassociation of distance from cost in information transfer, exemplified by the communication satellite, has extended the possibilities of transnational communication (Oettinger, 1980). The convergence of computer and telecommunication technologies, and the extensive demands placed upon U.S. and European telecommunication networks by corporate and government computer applications, have highlighted the importance of telephone networks. What were once plain old voice networks are now seen as the central conduits of the information society, analogous to the canals and railways of the industrial society (Goddard & Gillespie, 1986).

These developments have also served to highlight the tremendous inadequacies of the telecommunication infrastructures of Third World countries. As long as old-style PTTs (government post, telegraph, and telephone authorities) preside over unreliable local telephone networks connecting miniscule minorities of Third World populations, the promise of true transnational communication falls short of realization. The increased globalization of production processes and markets by transnational corporations (TNCs), and the expansion of trade linkages between Third World countries and the West, are hobbled. True integration of these economies requires the development of their telecommunication networks. Third World countries have also drawn the attention of highly rivalrous firms in the telecommunication technology field (and of their national governments) as markets with tremendous potential (Melody, 1985). The comments of Allen Greenberg (1985, pp. 42–44), "a senior Foreign Service officer associated with U.S telecommunications development affairs," exemplify the new attitude:

> Third World telecommunications are . . . of considerable interest to the United States, for reasons ranging from enlightened altruism to the bread-and-butter concerns of trade balances and jobs. The United States, with its huge internal market, advanced technology, and extensive aid programs, could be the world's leader in assisting the less developed countries (LDCs) in expanding and improving their telecommunications networks. Paradoxically and unfortunately, we lag far behind Western Europe and Japan in this effort. If present trends continue, the United States will soon be notable by its absence on the LDC telecommunications scene, with severe and probably irreversible damage to our domestic and foreign policy interests. . . .
>
> Since the United States now faces stiff domestic competition [in the telecommunications equipment market] and is shut out from the markets of the other major suppliers, a possible way to preserve the U.S. telecommunications equipment industry from going the way of shoes, color TVs, and cameras is to focus attention on the relatively open markets in the LDCs. At least this was the conclusion reached by the U.S. International Trade Commission in its 1984 report to the Congress.

There are domestic reasons for the rise of telecommunication technologies on the domestic agendas of Third World countries, too. As in the West, the spread of computer technologies has created demand for reliable data communications and other enhanced telecommunication services. This has been especially pronounced in the externally oriented sectors of Third World economies which must communicate with foreign countries at higher levels of information-communication technology applications (Sullivan, 1988). Military-security sectors, which take their cue from their more advanced counterparts in the West, are also more aware of the value of telecommunication facilities and constitute a significant lobby for advanced telecommunication technologies. Domestic transportation companies, administrative agencies, and domestic trade and industrial interests are among others demanding better telecommunication services. The long waiting lists for telephones found in most Third World countries testify to the middle class interest in voice telephony (Kaul, 1988). Industrial policy is an important, indirect factor in countries such as India and South Korea, where the widening of domestic markets for telecommunication equipment helps the growth and consolidation of domestic electronics industries (Brundenius & Göransson, 1985; see also Ambrose, 1988). Overloading a telecommunication network with connections (resulting from incremental responses to pent-up demand) and/or traffic (resulting from the abnormally high levels of usage characteristic of Third World countries) can dramatically degrade the quality of the entire network. This directly affects key decision makers and powerful interest groups and may contribute to the higher profile achieved by telecommunication in recent times.

In addition to the above-outlined domestic and international factors leading to the ascendence of telecommunication on the policy agendas of Third World countries, there exists an academic and policy discourse associated with the International Telecommunications Union (ITU) and the World Bank that focuses on the role of telecommunication in development (Hudson, 1984; *Information Telecommunications and Development*, 1986; Maitland, 1984; Saunders, Warford, & Wellenius, 1983). This literature explores the relation between telecommunication and socioeconomic development. In the context of rural development, much of the literature has been preoccupied with attempts to validate the hypothesis that telecommunication permits improved rural social service delivery, improved cost-benefits for rural economic activities, and a more equitable distribution of economic benefits (Hudson, 1987). Thus, rural telecommunication demand is also a factor in the policy arena. The interests of the potential beneficiaries may be articulated through their representatives in government or may be represented in the policy arena by third parties such as aid organizations or consultants. Or, it may be that rural telecommunication needs and development objectives serve other purposes, such as providing justification for resource allocation decisions taken on entirely different grounds.

Telecommunication resource allocation decisions may be seen in terms of supply decisions between investing in telecommunication as against other things; demand decisions where choices are made between classes of users; and another level of choice involving cost–price relationships. These decisions, be they taken by government or by private firms, involve choices that affect others in society, and hence are value-laden. Those who make these decisions are advised and informed by policy experts. The academic and policy discourse on telecommunication is not value-free, despite pretensions to the contrary (Fischer & Forester, 1987). This chapter seeks to lay bare some key value issues pertaining to telecommunication policy in the Third World.

Values are here understood as normative arguments, as judgements of good and bad. It is realized that there are no objective criteria for such judgements. What is argued here is that value choices are intrinsic to Third World telecommunication policy, and that it is better for these choices to be made in the open, and not surreptitiously smuggled in behind social scientific rhetoric.

CHOICES IN RESOURCE ALLOCATION

Telecommunication Versus No-Telecommunication

In general, the academic literature on telecommunication and development does not address the issue of telecommunication or no-telecommunication (*Information, Telecommunications and Development*, 1986, p. 10, pp. 38–40, are exceptions). Most writers assume that telecommunication channels are a good, and that communication linkages and information carried through them are beneficial to all concerned (e.g., Hudson, 1982, p. 163; Saunders et al., 1983, p. 19). What is left unstated is the strong likelihood that improved two-way communication between city and hinterland may result in the ruin of local traders in the short term, and the incorporation of hitherto relatively self-contained politicoeconomic entities into the metropolitan economy with a resultant loss of autonomy in the long term. Du Boff's (1983) study of the economic impacts of the use of the telegraph in 19th-century America, as well as Innis' (1956) analysis of communication-transportation channels in 18th-century Canada, are instructive in this regard.

At the very beginning of a discussion of telecommunication in the Third World one sees an unexamined value choice—an option for the value of integration of small or less-powerful units into large and powerful ones at the possible cost of local autonomy. But it may be argued that this value is inherent in telecommunication, or in any type of institutionalized communication for that matter. McQuail (1986, pp. 2–3) has asserted that the social scientific inquiry into communication, principally mass communication until now, has been influenced by a normative bias similar to the value of integration discussed here:

> the value of 'togetherness'—favouring community, solidarity, cooperation, integration, against isolation, fragmentation, individuation, 'privatization.' It is likely to be invoked in defence of established patterns of life and culture, of the specificity of language and belief. It translates into support for national, regional and local media forms and may also, less widely, be expressed in those forms of media which correspond to other bases of solidarity, such as class, religion or political allegiance. Related to this value (although sometimes viewed more simply as the reverse of freedom) is the value placed on *order*, in the sense of morality, tradition and continuity, as opposed to unregulated change and deviance from established standards.

The ambiguity of the definition reflects the inherent nature of the value. Integration necessarily involves simultaneous internal and external homogenizing thrusts: differences and deviance within the defined unit of integration are suppressed; the rest of the world is homogenized into a hostile "other." Two groups holding the identical value of integration may engage in bitter battle with each other, because their definitions of the unit of integration differ and overlap. Ethnic conflict, wherein the integrational thrust of nation states is met with the integrational thrust of ethnic movements, is an example.

Integration is dialectically related to the value of self-determination. The two values underlie all decisions affecting asymmetric power relations between countries as well as between groups and communities within countries. Communication channels, being carriers, not only of information, but also of power, are a crucial factor in the establishment, maintenance, and change of these power relations. Inasmuch as the Roman roads and England's control of sea lanes were essential for the maintenance of empire in earlier times, telecommunication is essential to the exercise of transnational and state power today. The corollary is that telecommunication is vital to the balancing of that power, namely the exercising of the rights of self-determination of countries, groups, and communities.

Institutionalized communication is made necessary by a fairly complex division of labor across space (Machlup, 1977). A spatial division of labor, by definition, reduces the autonomy of participating units, be they individuals or communities. Market relations are not at issue here. Spatial division of labor may be in a market context or within a planned economy. The different division of labor found within a self-contained, geographically compact entity such as a village (e.g., the caste system) does not require rapid communication over space. It is when the division of labor extends across space, eroding the self-sufficiency of the village community, that the need for communication becomes paramount. Of course, most Third World communities are no longer in states of pristine self-sufficiency. Thus, what are of real interest are communities that are relatively shielded from outside contacts. What are the implications of introducing telecommunication to such communities? What impact will there be on the existing internal communication networks, as well as on linkages with the outside?

The new must always be compared with the old, to reach an adequate understanding of the implications of change:

> Equally important—and totally overlooked in the great majority of studies—is an examination of the information and communication networks being used prior to the introduction of new technologies. Without knowing the prior information flows and communication relations, one has no base case against which to compare the new, changed relations resulting from the implementation of the new technologies. The failure to pay adequate attention to the base condition often results in a simple documentation of purported benefits of the new technology to those particular users who have benefitted. This approach . . . tends to draw the researcher into the role of myopic promoter of the technology. (Melody, 1986, p. 62)

It appears that many researchers in the area of telecommunication and development have fallen into this trap.

The issues posed by spatial division of labor within countries are also relevant to the international division of labor. The value issues of international communication have been discussed in the context of the debate over cultural imperialism (Schiller, 1976). What is surprising is the lack of articulation between international and domestic issues. Many writers who argue for complete or partial disassociation of Third World countries from world capitalism fail to work through the implications for self-determination for groups within countries.

Enforcing disassociation (e.g., establishing "cultural screens"—Smythe, 1981, p. 232) requires a strong state. Strong states strive to integrate groups within their domains into homogeneous units. Thus, with disassociationist states, there would tend to be a relatively higher emphasis on internal communication channels for purposes of integration and control. This would detract from the autonomy of groups within the nation-state. Centralization of power would go against local communities defining the content of development—an important element of many disassociationist writings (e.g., Goulet, 1979).

Market Versus Administrative Allocation

The market is a mechanism for allocating scarce economic resources. Economic resources may also be allocated by government or by monopoly firms through nonmarket, administrative means. Within a pure market context, satisfying demand for telecommunication services would not be considered a problem. Demand would be satisfied in the order preferences are revealed—those willing to pay the most will be served first by firms desirous of maximum profit. Other firms will be attracted to the market by the prospect of high profits gradually bringing down prices and matching supply with demand. At a facile level, the idealized market model does away with the need for value choices, since decision making is decentralized and each individual actor is assumed to have complete autonomy to act, either by expressing preferences for telecommunication or other economic goods in the role of consumer, or by deciding which market to supply in the role of producer. Private interests are pursued with no thought for the interest of others. The market model appears to provide an assurance that the interests of others are looked after as though by the working of an invisible hand, minimizing if not eliminating the importance of value choices affecting others.

> It is not from the benevolence of the butcher, the brewer or the baker, that we expect our dinner but from their regard to their own interest. We address ourselves, not to their humanity but to their self-love, and never talk to them of our own necessities but of their advantages. Nobody but a beggar chooses to depend chiefly upon the benevolence of his fellow citizens. (Smith, 1982, p. 119)

The construction of a plausible system of resource allocation on this apparently value-free base is a principal attraction of the market mechanism.

Yet the very choice of the market mechanism is a fundamental value choice. The theoretical model of the market corresponds to a level playing field where all have equal opportunity. The fact that some win and some lose in the game is justified because equal opportunity, a very modest version of the value of equality, is said to exist (Schaar, 1967). But in the real world, the playing field is not level. Not everyone has the dollars to vote with or take entrepreneurial risks with. The choice of the market as a resource allocation mechanism offers no safe harbor from value decisions. It is not that the polar opposite of the market, pure administrative allocation, offers a better alternative. Resource allocation by purely administrative means is problematic, as evidence most recently by the self-criticism offered by countries with the greatest experience of administrative resource allocation, the Soviet Union and the People's Republic of China.

It is well known that all markets are imperfect. There is agreement across the ideological spectrum—from libertarians (e.g., Hayek, 1944; Buchanan & Tullock, 1962), from welfare economists and liberal economists (e.g., Scherer, 1980), from institutional economists (e.g., Commons, 1924), and from radical economists (e.g., Cole, Cameron, & Edwards, 1983) on this point. The real issue of contention is whether, and under what conditions, administrative al-location is superior to an imperfect market. One of the most important causes of market imperfection is lack of information, the inadequacy of the assumption of perfect knowledge. For a market to work well, participants must have knowledge of the economic and technical data relevant to their decision making. Markets in market information, wherein access to such data may be granted or denied on the basis of ability to pay, are not conducive to the gaining of that knowledge by all parties. In any case information is not easily commoditized. Imperfections in information markets will cause and exacerbate imperfections in markets for noninformation goods (Melody, 1981; Newman, 1976). Thus it appears that provision of market information through nonmarket mechanisms may be a re-quirement of perfect markets in noninformation goods. However, the task of devising effective nonmarket mechanisms for information provision is a difficult one (Arrow, 1971). Given that difficulty and the ideological commitments of most economists, the issue has become one of incremental improvement of markets through better provision of information. The discussion in the literature about telecommunication improving the efficiency of Third World product and factor markets must be located within this context (e.g., Leff, 1984).

But what of the choice of market or administrative means for allocation of telecommunication resources? The issue is not simply one of the relative effi-ciency of one mechanism over the other. It may be that a social choice will be made to allocate resources to telecommunication over and above what the market may allocate in order to improve the functioning of other markets.

In actual fact, telecommunication decision makers in the Third World func-tion by and large within the public sector (a domain of administrative alloca-tion) and not within the market. This is partly due to the inherited institutional structures of PTTs as government departments. But the most important factor is the capital-intensive, infrastructural character of telecommunication facilities. The weak capital formation mechanisms found in Third World countries are incapable of investing in projects which require major upfront commitments of capital and long gestation periods. There are significant constraints on access to international capital markets (Goldschmidt, 1984). Telecommunication networks are characterized by significant externalities. Among those mentioned in the literature are declining average costs over a wide range; increase in the welfare of existing subscribers as new participants join the system (more people to whom calls may be made through the network); and improvement of the functioning of markets and organizations through the provision of better and more current information (Leff, 1984). When significant externalities exist, the market is not the optimal resource allocation mechanism. Policy has a role in intervening with nonmarket instruments or in designing property rights for the effective functioning of markets (Reich, 1987, p. 224; Samarajiva, 1985, p. 167).

There is general consensus in the literature about the legitimacy of non-market resource allocation mechanisms for telecommunication in the Third World. Policy makers as well as policy analysts recognize that Third World telecommunication development necessarily involves decisions regarding the allocation of public resources (i.e., public funds, loans obtained on government surety, tax concessions, protection from competition). These decisions pose value choices.

Telecommunication Versus Alternative Investments

Value choices made in the allocation of public resources to telecommunication facilities as against other things such as food for the hungry, infrastructural projects such as irrigation or roads, subsidies for industries, armaments, casinos, or whatever, are extremely important. The literature does not address this choice in an adequate manner. This may be because those who engage in the academic and policy discourse are already committed to telecommunication through sunk costs of education, job choice, etc., and have no incentive to critically examine the benefits of investing public funds in telecommunication as opposed to other things. The decision to invest public resources in telecom-munication tends to be discussed as a question of empirically establishing a positive contribution from telecommunication to development (e.g., Hardy, 1980), or as a problem of disseminating the finding of a positive contribution to the right audiences (e.g., Parker, 1984; Wellenius, 1984). Even if one were unquestioningly to accept the proposition that telecommunication makes a positive contribution to development as measured by GNP, that would have marginal relevance to the decision to allocate public resources to telecom-munication as opposed to other things. That decision must rest on opportunity cost: Will a given amount of public funds yield greater output (in whatever terms development is measured) if invested in telecommunication rather than in rural roads, school meals, casinos, or whatever? This is the correct question, but it is one that is exceedingly difficult to answer. This question is not asked, let alone answered. The basic decisions on telecommunication resource allocation are not taken on the basis of "hard, value-free" empirical evidence. In reality, values play a central role in decision making at this level.

In any case, empirical evidence generated by cost–benefit analysis, input–output analysis, and correlation studies is not value-free. The choices of what factors are to be examined, quantified, and correlated rest on value assumptions (Fischer & Forester 1987). Even more problematic is the underlying notion, pervasive in policy analysis, that the state is a benevolent body that will act upon rational evidence for the greater good of society.

Given the absence of data for rational administrative allocation, can the decision maker seek assistance from market signals? Can the government behave like a capitalist in investing resources in products and services for which there is demonstrable demand in the form of customers willing to pay? This is the course of action prescribed by the World Bank and its experts (Saunders, 1982, p. 205).

The question has three parts: Should government invest public resources in telecommunication as against other things on the basis of market demand? Once telecommunication investment has been decided upon, which user groups are to be served? And what relation will there be between the costs of providing telecommunication services and the prices charged specific user groups? Not knowing the opportunity costs, the first part of the question cannot be answered. Yet decision makers do decide to invest in telecommunication all over the world. It must be that these decisions are based on value judgements. The second part of the question presupposes that the telecommunication investment decision has been taken. The question then is, who gets the phones or the data links? Should government take this decision on the basis of market signals, as proposed by the World Bank?

Public resources belong, by definition, to the public: all the people. One may justify unequal allocation of public resources when the resource flow is towards the poor, in that society as a whole has an interest in reducing income disparities based on the values of equity, equality, and participation, manifested in forms such as providing basic needs, leveling the playing field, or not allowing the marginalization of groups in society. Society may also justify resource transfers to the poor in the interest of forestalling revolt and insurrection. It has been argued that the provision of basic economic rights such as telephone service is part of the social perception of economic justice, even in market-oriented countries such as the U.S., and that perceptions of justice strongly influence economic policy (Zajac, 1981, p. 102). However, one would be hard put to justify the unequal flow of public resources to the rich, which is what directly allocating funds for the provision of telecommunication services for those most willing to pay would amount to. Of course, the argument is made that such investment is beneficial to the poor in the long run because the monopoly profits can then be ploughed back in the form of investments directly beneficial to the poor:

> The results of such an analysis in those developing countries in which there are large telephone and telex waiting lists and system call traffic congestion during daytime business hours will be to point in the short run to a pricing and investment strategy which includes relatively high connection fees and monthly rentals from

urban area subscribers and accompanying high busy hour call charges. The resulting increase in revenue could assist in financing increased expansion of subscriber
telephones in urban areas and help finance a significant "public access" program
designed to provide public call offices in urban slums, unserved towns, and remote
provincial villages. (Saunders, 1982, p. 192)

But the writer is honest enough to add that, "if government priorities dictate,
some of the surplus local currency revenue might be made available to government for general expenditure purposes." The postscript better reflects the reality,
albeit partially. The coffers of Third World governments benefit from monopoly
pricing, not rural populations.

The proposed method of providing telecommunication services to those who
cannot make their demand known through market signals is convoluted and
unrealistic. It may be described uncharitably as a smokescreen behind which the
rich grab the goodies. If monopoly profits can be made from the provision of
telecommunication services to the rich, it would be far better for government to
open up that market for private entrepreneurs and tax their profits. If entrepreneurs refuse to enter such a market without the safeguard of a government-
enforced monopoly, it may indicate either that they do not see the potential for
high profits or that the risks are too high. In either case, the rationale for
investment of public funds would collapse.

The World Bank arguments may not be water-tight, but they do reflect what
is actually happening in Third World countries. Given the absence of any
empirical basis for allocating public funds for telecommunication, government
decision makers have been responding to a melange of market signals and
stakeholder demands. Long waiting lists, congestion during peak periods, and
willingness to pay exorbitant connection fees and rentals do exist and constitute
strident market signals. TNCs and export oriented industries make strong demands for telecommunication services, as do administrative and security agencies. The groups thus indicating their preferences are also those capable of
exerting direct pressure on government decision makers. Government decision
makers who have found themselves in the position of monopoly providers of
telecommunication services have responded by allocating the majority of telecommunication funds to satisfy the needs of those making their demands known
(Clippinger, 1977).

The magnitude of the monopoly rents being extracted have obscured both
the tremendous inefficiencies of the PTTs and the extent of utilization of public
resources in the provision of telecommunication services to a few. It has been
observed that most Third World PTTs are profitable and that they contribute to
the general treasury (Saunders et al., 1983, pp. 12–14). This appears to run
counter to the proposition advanced above that public funds are utilized for the
provision of telecommunication services. Yet it is also a fact that government
funds, foreign exchange allocations, and/or credit guarantees are referred to by

all writers on the subject of telecommunication investments (Carreon, 1976, p. 128; Goldschmidt, 1984, p. 183; Saunders et al., 1983, p. 16). This suggests, at the minimum, that the direction of the net flow of resources is problematic and deserves to be investigated in detail. It is likely that a true accounting would show a long-term net outflow to the PTTs, given their tremendous inefficiencies. In the short term, the facts remain that capital investments in telecommunication take away funds that could be used elsewhere, and that the application of the value criteria of equity and equality is not inappropriate.

It is, however, possible to envisage a scenario where telecommunication resources can be allocated to specific networks such as data communication networks to be used by clearly identified user groups such as export industries, with provision made for complete cost recovery. It may be that clear externality benefits or public policy objectives necessitate such action. Usually, what happens is that other telecommunication users are made to cross-subsidize such specialized networks. But clear separation, on the lines of the separate subsidiary concept in U.S. telecommunication, may allow for flexibility without cross-subsidy.

Production Versus Consumption

This section examines the value implications of the allocation of telecommunication resources on the basis of contribution to productive activity (i.e., improvement of the coordination and organization of economic and administrative activities) or to consumption (i.e., contribution to quality of life by facilitating communications with kin and friends and providing access to emergency and other services, as is the case in North America). The question of whether telecommunication should be used primarily as a productive input or as a consumption item must be situated in the wider context of the development debate of the past three decades.

This debate revolves around the question of whether resources should be allocated to productive uses in order to accelerate growth, thereby raising incomes and the living standards of the poor, or whether resources should be allocated directly to consumption—the provision of food, education, housing, and health. The latter necessarily involves redistribution between different groups unless the financing comes from external sources (Bhagwati, 1988).

The dominant theory of economic development during the 1950s and 1960s was a theory of growth. In this theory, the primary focus was placed upon capital accumulation and investment. Mobilization of savings was characterized as the key to generating productive investment (Lewis, 1955). In this growth strategy, the state was assigned an interventionist role. It was to mobilize internal and external savings and determine investment priorities. Guided by this "conventional wisdom," the United Nations' First Development Decade (1960–1970) set a target of a 5% annual growth rate for Third World nations. Industrialization was seen as the principal instrument for achieving this objective. Governments allocated resources to the construction of hydroelectric dams, roads, factories, etc. It was hoped that this infrastructure would facilitate increased economic activity, enhance productivity and increase employment opportunities, thereby raising the incomes and living standards of the poor.

Subsequent experience revealed the weaknesses of the indirect growth-based approach to the amelioration of poverty. The United Nations First Development Decade was a success in terms of growth rates. However, the poor became poorer in many Third World nations (ILO, quoted in Etienne, 1982). New thinking within academic circles and international organizations such as the ILO deemphasized the restructuring of patterns of production and income and focused more on the consumption side of the economic equation. Resources were to be allocated directly to satisfying the basic needs of the poor: the provision of food, health services, sanitation, and education.

There are several strands within the basic needs approach to development. They diverge on how much resources should be allocated directly to consumption and on the role of the market in this process. The redistribution of growth approach advocated by Chenery et al. (1974) argues that, in addition to allocating substantial resources to productive pursuits, all members of society must be provided with adequate income to participate in a market system. This income will enable the poor to satisfy their basic needs and for markets to develop. The approach of Streeten, Burki, Haq, Hicks, and Stewart (1981) also looks at both production and consumption. It is accepted that absolute poverty can be eliminated permanently only by increasing the productivity of the poor, but it is also argued that these efforts must be supplemented by steps to satisfy their unmet basic needs. It is stressed that basic needs is not primarily a welfare concept: Improved education and wealth can make a major contribution to increased productivity. Indeed, this "human capital" argument is the defense used by Streeten against the criticism that there is a tradeoff between growth and basic needs. The outcome of this debate is still unclear. Streeten disagrees with Chenery on the point that all basic needs can be satisfied in the marketplace. One reason is that the incomes of the poor are too low. Another is that producers will not adequately invest in the supply of basic needs, concentrating instead on other areas yielding better financial returns (Cole & Lucas, 1982). Thus, basic needs have to be provided directly through government programs. A final strand of the basic needs approach asserts that growth is irrelevant in the battle against poverty. The growth/basic needs tradeoff is considered a red herring. Government should direct its resources to redistribution and satisfying the basic needs of the rural poor (Griffin, 1981).

Historically, telecommunication in the Third World has been used in a productive sense (i.e., contributing to coordination and organization) by government, military-security, and business interests and, to a lesser extent, as a consumption good by the urban elites. For example, in Papua New Guinea, the telecommunication system was found to be heavily biased in favor of serving the productive needs of the export-oriented enclave economy. The average telephone density for Papua New Guinea in the late 1970s was 1.3%. However, only 0.6% of the total indigenous population of Papua New Guinea had telephones. Over 30% of expatriates (engaged in managing plantation and mining industries) had telephones (Karunaratne, 1982). The unqualified use of aggregate telephone density figures for urban and rural areas obscures the true determinants of telecommunication resource allocation. Costly telecommunication lines are drawn to the most inaccessible rural areas to serve plantation and mining industries; the urban poor and lower middle classes (clerks and teachers) in the urban areas are excluded from access to residential telephones.

The human capital approach brings out the difficulties of defining productive and consumptive goods (Blaug, 1976; Johnston, 1977). Is education a consumption item or a productive input? Is improved healthcare for workers (cutting down absenteeism and improving productivity) a consumption item? In the same way, it is extremely difficult to clearly demarcate the line between productive and consumptive uses of telecommunication. This is especially the case with small and emergent businesses and professionals. Even more problematic, but still important, are questions such as the characterization of the provision of telecommunication services for seasonal migrant workers who can use the services to keep in contact with families. Such links may also serve to improve the functioning of the labor market. The provision of telecommunication services to urban slums which can be used to facilitate the intense entrepreneurial activities—legal and, sometimes, illegal—of the urban poor are another example.

Leaving it to the market to allocate telecommunication resources to production or to consumption as recommended by Saunders (1982) would simply reinforce the historic patterns of telecommunication facility distribution. Emergent or small businesses, groups such as lower-ranked salaried employees (e.g., clerks and teachers), and the rural and urban poor will be excluded from telecommunication services (for production and/or consumption). Would this not be a value choice against equity and equality, especially since public resources are being expended?

Writers such as Parker (1982, 1984) and Leff (1984) who advocate administrative allocation of telecommunication resources build their case on productive use. Parker sees information, and its carrier telecommunication, as a vital productive input into the economy. The telecommunication infrastructure is seen as performing an essential function by facilitating the flow of information about improved products and techniques in the industrial sector, and in the coordination of large-scale economic projects. In the agricultural sector, improved access to information via reliable telecommunication is seen as having the potential to improve agricultural productivity by efficiently disseminating research findings on improved seeds and fertilizers, and by providing market information. Parker also sees telecommunication as an input to the administrative machinery delivering development programs to rural areas. Telecommunication is a key administrative tool in providing consumption items to the rural poor—health, education, and other services. Parker (1984, p. 174) advocates, "not just more urban residential telephones but, more importantly, the provision of [telecommunications] . . . to every location where significant economic activities are planned."

The justification for Parker's three-point advocacy (telecommunication for industry, agriculture, and administration) derives from his position on balanced growth and basic needs that has much in common with the approach of Streeten et al. He points out that successful industrial development strategies can lead to urban migration and consequent deterioration of quality of life. Compensatory programs are thus justified. However, if rural life is improved only by basic needs programs without rural productivity gains, the rural sector will become a drag on the rest of the economy. This is where rural telecommunication is seen as capable of playing a vital role (Parker, 1976).

Parker assumes that integrating economic activities in the rural areas into the national economy is beneficial to all concerned and that improved access to information will actually make a difference. Yet these assumptions are questionable. The first involves a value choice in favor of integration, as discussed above. Parker is right in arguing that the market is a poor indicator of demand for rural telecommunication, in that, in many rural areas, there is no intersection of supply and demand, and hence no supply at all (1982, p. 3). But does that make him the sole judge of what those needs and demands are? This raises the issue of participation in defining and assigning priorities to needs (Clippinger, 1977; Huntington & Nelson, 1976; McAnany, 1980). Even if government officials, academics, and consultants know what is "good" for these users, will they (or can they) act upon those valuations, given their location within power structures? Participation is a value that touches on the values of equity, equality, and integration.

Parker's second assumption has been questioned by studies such as O'Sullivan's (1980) which suggest that there are situations where improved access to information makes no difference. One may be "throwing" information at problems that are rooted in structural constraints. Current thinking on communication and development posits the requirement of a careful examination of the social, political, and information environment of intended beneficiaries. This is necessary in order to prevent the waste of resources, the dangers of reinforcing existing exploitative relationships, and the possibility that current channels of communication and exchange may be undermined without being replaced.

Hudson, the leading writer on telecommunication and development, and organizations such as the ITU, subscribe to most of Parker's arguments. However, in addition to advocating telecommunication as a productive tool, they make some mention of its role as a consumptive good for rural areas. For example, it is pointed out that rural dwellers can use telecommunication facilities to keep in touch with family and friends and to reduce the sense of isolation. It reduces the need for travel, even for those not directly involved in productive activities. The ITU's proposal for a telephone within an hour's walk of the majority of rural residents arose from this context (Maitland, 1984). Hudson (1984, pp. 30–31) also makes reference to the political uses of telecommunication, pointing to the possibility of two-way communication links between urban and rural areas making it easier for rural people to make their demands known and become more effective in claiming their "fair" share of national budgets.

However, closer inspection of her writing makes it clear that consumption is not for everyone:

> If people perceive themselves as clusters of families with no links between them except for blood ties, they have not reached a point of collective identity where organization is likely to take place, and thus will not use media as tools needed for organization. (Hudson, 1984, p. 26)

According to her typology, individuals in such "clusters" belong to Type I; they have no understanding of their external environment and will use media only for "personal" reasons (presumably this means consumption or perhaps local trading). The prescription is that Type I individuals should not be given priority for telecommunication resource allocation. Priority should be given to Type IV individuals—the leaders and representatives of a large or scattered constituency, with knowledge of the external environment, who will engage in seeking information for collective goals. These individuals will tend to be the articulating elements in the local society: political leaders, merchants, agricultural producers with surpluses for sale outside the community. It thus appears that, when it comes to the crunch of resource allocation, Hudson's advocacy of telecommunication as a consumptive good blends into her larger assumptions of integrated, market-driven economic models. The real beneficiaries of resource allocation in telecommunication in the rural areas are not the rural poor, but the articulating elements that mediate between them and the urban economy. Despite Hudson's disclaimer (1984, p. 16), it appears that "trickle-down" is part of her theory.

Not deciding between productive and consumptive uses, and allowing the market to decide allocation, will reinforce existing patterns of telecommunication service distribution, though there may be a shift from government officials to business and professional users within the urban elite. The approach advocated by Parker and Hudson emphasizes productive use and would favor government officials located in rural areas and the articulating elements of rural communities. This is based on value choices that favor the integration of rural communities and their economies into the metropolitan economy, and detract from the rights of such communities to define their development priorities.

VALUES IN POLICY: AN EXAMPLE

This section illustrates the importance of examining value issues in policy, in relation to choice between telecommunication technologies that connect geographically proximate points (intracommunity linkages) and those that connect distant points (intercommunity linkages, usually metropolitan–hinterland linkages). It must be noted that this discussion in and of itself does not constitute an endorsement of a particular technology. Such conclusions can only be drawn by institutional analysis of the particular setting into which the technologies will be introduced.

Most telecommunication usage occurs within geographically compact areas (Mayer, 1977). Deviations from this pattern exist only in a few areas such as Alaska and Northern Canada which have exceptional demographic and geographic characteristics (Hudson, 1984, p. 32). Historically, telephone service evolved in North America first as a local service, with toll or long-distance service being added later (Gabel, 1983). The present controversy in North America over cross subsidies and universal service has been triggered by the perceived shift of a part of the cost burden of the telecommunication system to the majority of the population that makes few or no long-distance calls, from heavy long-distance users (primarily business and government, and a minority of residential users) (Weinhaus & Oettinger, 1988).

Therefore, it may be surmised that local service would be given primacy in Third World telecommunication planning. This does not appear to be the case. Most writings about the revolutionary possibilities for rural telecommunication opened up by satellite technology assume the centrality of the need to connect rural points to the city (e.g., Abramson, 1984; Casey-Stahmer & Goldschmidt, 1985; Hudson, 1984; Parker, 1982, 1987). The choice of technology has been framed as one between satellite versus terrestrial technologies such as high-frequency radio and microwave, the flexibility of the satellite versus the rigidity of the "wired tree." The choice between technologies primarily conducive to local communication, and technologies that link rural communities to the metropolitan centers to the possible detriment of local networks, is not considered worth examining.

The value choice favoring integration is routinely made by Third World governments as well as by writers on the subject. Most Third World states tend to be of the centralizing type. The dominant development models, be they drawn from development economics or from the more recent disassociationist approaches, assign a preeminent role to the state and do not assign major decision making roles to local communities. Even the basic needs approach essentially envisions the inputs for the achievement of basic needs coming from the city (or the Western nations) to the rural areas (Bhagwati, 1988). In this context, it is not surprising that all the useful information is thought to be located in the city and that the only useful communication links are those connecting rural areas to the metropolis. This thinking also meshes well with the security imperatives of the Third World state, a key aspect of telecommunication policy understated by officials for obvious reasons and ignored by scholars for inexplicable reasons. The following excerpt from a statement made by the head of state of Senegal at the inauguration of a new stretch of the Panafrican telecommunication network illustrates the high value ascribed to security needs in relation to telecommunication systems:

> All problems relating to [telecommunications] are examined from the triple standpoint of security, strategy and sovereignty. . . . Telecommunications development advantages . . . might be said to cover three essential functions:
>
> - a security function,
> - a social function, and
> - an economic function.
>
> . . . The role of communications remain paramount in the effort to achieve national and regional economic integration. (*Information Telecommunications and Development*, 1986, p. 41)

The short-term objectives of Third World governments in relation to telecommunication are clear. What is not clear is the position almost uniformly adopted by the writers on the subject. On first glance, the fact that researchers most active in the telecommunication and development area, such as Hudson, Goldschmidt, and Parker, come from a background of involvement in telecommunication issues in Alaska and the Canadian North (Hudson, Goldschmidt, Parker, & Hardy, 1979) may appear to explain the emphasis placed on distant communication. These areas are extremely sparsely populated, with resource extracting camps and small communities of natives scattered across harsh terrain. Both these groups are very strongly integrated to the metropolitan centers. Resource extraction is carried on by those from the metropolis for the benefit of the metropolis. The need for extensive links with the urban areas is obvious. The situation with the native communities is more complex. These groups are survivors of a wrenching social transformation and as a result are quite dependent on support payments, food, and other resources from the urban centers. Recent land claim settlements have improved the politicoeconomic positions of some of these groups but tie them to the resource economy dominated by the urban centers. The extensive use of long distance telecommunication in Alaska and Northern Canada arises from these specific circumstances. Parker (1981) completed an extensive study of telecommunication systems serving rural populations in the continental U.S. and should have been aware of the different patterns of telephone usage found in those areas. It would not be reasonable to generalize from such communication patterns to the rural areas of Third World countries which are at quite different levels of integration to city-based market economies and do not, as a rule, share the demographic characteristics of Alaska and the Canadian North. Furthermore, it is surprising that none of these writers refer to the security dimensions of rural telecommunication, given the fact that the basic telecommunication infrastructure in Alaska and Northern Canada was constructed for military purposes and was managed by the military until the end of the 1960s (Mansell, 1979).

The preoccupation with public call offices (PCOs) (Maitland, 1984) is also part of this rural–urban fixation. The PCO or pay-phone technology is designed for calling out. It is unsuitable for true two-way communication. What it is appropriate for is for phoneless people to initiate communication with those with residential or business phones. This perfectly fits into the asymmetric model of rural–urban communication implicit in government policy and academic writings.

The installation in the rural areas of a technology designed for urban–rural linkages has not only been promoted unquestioningly, but the researchers have also tried to come up with theoretical explanations for the ensuing usage patterns. Hudson (1984, pp. 26–28) classifies rural people into types á la Lerner, the leading communication and development theorist of the 1950s: Type I persons, whose relationships are primarily local are seen as "likely to use the media only for personal reasons"; Type III persons who are more outward oriented are said to be at a take-off point; and the Type IV person, the ideal type who will effectively and enthusiastically communicate with the "outside world" or the metropolis where the telephone line ends (Type II is described as "logically a rare condition"). Is it not obvious that the technology has been designed for Type IVs? Is this not a case of confusing the symptom for the cause? If a telecommunication system that connected all the rural people in the community (as was the case with early North American rural systems) were to be established, one could expect the Type Is to become the innovators and some of the Type IVs to become laggards.

The implications of the value choice favoring integration may be seen in the following conclusion by Hudson (1982, p. 167):

> An extremely decentralized and autonomous system may benefit little from a telecommunications network. Interesting examples are two central African countries which requested a loan to extend and improve telephone services. The question of the benefits of including rural public call offices was raised. Such benefits are extremely difficult to predict, as the economies are predominantly rural, with farmers generally living on small individual land holdings of one or two acres scattered throughout the hilly countryside. There are almost no villages. The government plans call for regional development centers to provide services and start local industries. If and when these centers materialize, it could be argued that rural telecommunications could be used to provide market information, order supplies, and provide administrative links between these poles and regional and

national centers. In the meantime, it would appear that rural telecommunications could best be planned to serve rural industries such as tea and coffee plantations and rural institutions such as hospitals and residential schools (many of which now have dedicated HF radio systems).

The value assumption leads to a specific technology choice, which then biases the allocation of resources to the sectors of the rural economy integrated to the urban-based market system.

It is not that the distant bias is inherent in all telecommunication technologies. The rural digital exchange, a state-of-the-art technology, has potential for both distant and local communication. Basically an upgraded version of the PABX (private automated branch exchange) developed for the office market in developed countries, rural digital exchanges are microprocessor based, compact, reliable, and relatively cheap. Some may be overengineered for basic rural services in that they have the capacity to handle data and voice and because they have fairly sophisticated accounting and management capabilities. Yet the amortization of significant portions of R&D costs by the office market, large-volume production, and the competitive nature of the PABX and small rural exchange markets makes these technologies economically attractive. This is a technology in which the suppliers are not limited to the standard Japanese, North American, and European firms. Firms from India, South Korea, and Brazil are among those manufacturing rural digital exchanges, and some have been quite successful in gaining market share (Ambrose, 1988).

Some PTTs, such the Indian PTT, plan to locate rural digital exchanges in local shops to be administered by shopkeepers. The equipment is so compact and rugged that this is a practical possibility. The microprocessor-based design enables easy maintenance with replacement of circuit boards by personnel who need not be highly skilled. Should the exchange be linked to an urban center, it is possible to utilize the remote diagnostic capabilities too. However, it must be emphasized that the question of which social groups actually get to utilize these facilities is determined by the second level choices made regarding cost–price relationships, and not by the technology itself.

The rural digital exchange is first and foremost a local communication technology. It can provide effective two-way telecommunication services to local people who have things to talk about. In addition, it has the capability of linking those people to the national system if desired. This is the case with almost all the installations. However, the rural digital exchange is quite unlike the highly touted star-configured VSAT (very small aperture terminals) network. A VSAT network comprises of a sophisticated, and expensive, master antenna or hub and inexpensive terminals located in user premises. The more terminals there are, the more economical the system becomes. Most VSAT networks are designed for point-to-multipoint data transmission. Here, cheap receive-only terminals are served by the broadcasting master antenna. In the interactive configuration, more expensive interactive terminals can communicate with the master antenna too (for details, see Parker, 1987). The first major interactive VSAT network in a Third World country, NICNET in India, is of this type. NICNET is used to transmit government statistical data to and from regional centers to the central depositary in New Delhi (Blair, 1988). Individual users in interactive VSAT networks can easily communicate with the master antenna, but communication between users requires a 90,000-mile-long double-hop through the hub, which more or less rules out voice communication, a point conceded even by its most avid promoters (Parker & Rinde, 1988).

The key feature of the rural digital exchange is its openness to different applications. It will provide effective local communication, but with its queuing and other capabilities it can also utilize the trunk lines to the outside (the scarce resource) efficiently. It can provide basic voice, but it can also very easily be used as a messaging system (Gupta & Ramani, 1981). This can alleviate some of the problems caused by shortage of telephone instruments and "telephone tag." This type of messaging system will not be an exotic frill, but an upgraded version of telegraph service, highly popular and heavily used in Third World countries. The only additional work that will be required for an effective messaging system will be the incorporation of the ability to handle non-Roman scripts.

The most fundamental value issue pertaining to telecommunication in the Third World is integration/self-determination. Integration is at the heart of the choice between local and distant telecommunication technologies. One need not accept disassociation as a viable option for Third World countries or for communities within them to appreciate the importance of the issue in telecommunication resource allocation decisions. One need only recognize the asymmetric nature of power relations between metropolis and hinterland at the international and national levels. Communication technologies can easily accentuate those asymmetries. If the rights of small and relatively powerless groups and communities to exercise some control of their destinies are accepted (this is obviously a value choice), then telecommunication resource allocation must contribute to the strengthening of local communication networks along with the establishment of links with the outside. The key is balance between local and external networks.

Maintaining this balance will contribute to the realization of the values of participation, equity, and equality. Participation is better served by strengthening local networks, enabling people in small communities to talk to each other easily, rather than by artificially creating privileged interlocutors through the institution of unequal and biased patterns of access to communication channels. If the local networks of communities are strengthened, it will be possible for them to participate in wider politicoeconomic relationships more effectively. This type of participation is crucial to the realization of the values of equality and equity:

Socio-economic development and participation, do, in large measure, go together, but the connection between them is more complex and ambiguous than it is often assumed to be. Equality and participation are closely connected but the causal flow seems to be as much from the latter to the former (as posited by the other models) as from the former to the latter as by the liberal model. (Huntington & Nelson, 1976, p. 27)

The establishment of telecommunication services primarily serving to connect articulating individuals from rural areas to the metropolis will result in the weakening of existing local networks and the creation of new networks in which those individuals function as nodes. The earlier existing local market relations will be disrupted, since production will become oriented to the requirements of distant metropolitan markets and local products will face competition from distant suppliers. There will be good effects in that certain monopolies will be destroyed. Yet it would be unrealistic to assume that there will be no negative effects. The integration of an undeveloped and localized market to a larger and more developed market will most probably result in disproportionate benefits flowing to the more experienced and better capitalized players in the metropolis.

It is possible to visualize an alternative scenario. Here, the first step would be a study of the networks and economic relationships of the rural area, as well as telecommunication needs, since value choices have been made to favor self-determination and participation. Most probably, the study will also show that there is some demand for telecommunication linkages to the outside—the metropolis as well as certain other rural areas (e.g., a source of seasonal employment). A locally oriented telecommunication technology such as a rural digital exchange can be established, and most of the available resources expended on providing facilities for the enhancement of existing local networks, even at the cost of delaying or skimping on the external link (i.e., providing only a telegraph-type messaging service that is said to be cheaper than a voice channel, Abramson, 1984). Efforts are directed to the improvement of the functioning of existing local markets (i.e., village fairs, farmers' markets) rather than immediately connect rural producers and merchants to the larger and more developed metropolitan market.

In this approach, local networks are strengthened parallel to, or even prior to, new external networks. The objective is not disassociation but the building up of the capability to associate without being dominated. The use of the new technology in the local network first will serve a number of purposes. There will be less of a need to promote the use of the new service. The telecommunication network will connect people who know each other, have dealings with each other, and have things to talk about. Local use will familiarize the rural users with the technology, so that it will be transparent to them by the time they begin to utilize it for outside dealings. Local usage will also build up familiarity with telecommunication, and therefore demand. This will enable the use of cost-saving technologies such as multiplexing on the external lines.

Locally oriented telecommunication networks can enhance political participation. Hudson's (1984, pp. 30–31) vision is of individuals petitioning the powers that be in the metropolis. The strengthening of local networks may make it possible to build stronger local organizations which will be in much better positions to negotiate with metropolitan decision makers. Such participation will serve to realize the values of equity and equality.

Underlying the bias towards integration in the literature is a faulty communication model. Communication is seen as a power-neutral, interactive process in which the source and receiver share responsibility for what happens. This form of communication is thought to be facilitated by two-way communication technologies. But this does not correspond to what actually exists in society. The "communication as dialogue" which is assumed is often incompatible with existing social structures.

Many communication scholars who deal with two-way communication technologies tend to stress the interdependency of communicational relationships but fail to make allowance for dependency relationships within the context of interdependency. This point that has been made by Mansell (1982, p. 53) with regard to relationships between Third World countries and developed countries is equally applicable to relationships within Third World countries.

CONCLUDING REMARKS

Examination of a number of key resource allocation choices in Third World telecommunication development has revealed the importance of the underlying value choices. The authors of this article favor the values of self-determination, equality, equity, and participation. Yet, what is of importance is not whether everyone participating in the telecommunication resource allocation process and the associated policy-academic discourse adheres to these particular values as defined by us. That would be almost impossible, given the power relations within which the allocation decisions are taken and the policy-academic discourse is conducted.

Our plea is for an open discussion of values, whatever they may be. This plea, as everything else, is based on a value—that of participation. Clippinger (1977), who seriously examined the benefits of telecommunication projects to Third World peoples, called for broader participation in resource allocation decisions but expressed concern about the difficulties posed by participation in complex technological and investment decisions. Our conclusion is that popular participation that is not "anarchistic and . . . disruptive" can be ensured by making explicit the value issues underlying the most complex technological and investment issues, as we have attempted to do, and opening them up for discussion. For optimal effectiveness, the laying out of the value issues and the provision of opportunities for open and informed discussion must be institutionalized as part of the planning process, and the technical planning phase must incorporate the conclusions of the value discussions. These actions must be taken by telecommunication policy makers.

We recognize the idealism of the above proposal. A realistic appraisal of power relations within the Third World state (and international organizations) does not leave much hope that decision makers will cede power voluntarily in this manner, even though there is some evidence that such action may be in their interests (Amy, 1987). Much of the responsibility for introducing participation to the planning process and making considered value judgements devolves upon those who are called upon to assist in the actual planning as experts—the academics and the policy practitioners. The choice of methods lies in their province. Arguments may be marshalled for appeals to the enlightened self-interest of the power holders. The dissemination of understandably written policy papers and scholarly and popular articles which lay out the value issues may be another course of action. It may be necessary in some instances to adopt explicit advocacy roles, within government or without, in association with factions within government bodies or with outside interest groups.

The very recognition of the value-based nature of social science research places a responsibility upon its practitioners. If value choices are being made in research, there is a responsibility to strive to make the right choices and to have them reflected in the policy process.

REFERENCES

Abramson, N. (1984, March). Satellite data networks for national development. *Telecommunications Policy*, pp. 15–28.

Ambrose, W. W. (1988). Rural telecom: Growing market for digital exchanges. In D. J. Wedemeyer & M. R. Ogden (Eds.), *Telecommunications and Pacific development: Alternatives for the next decade* (pp. 289–293). Honolulu: Pacific Telecommunications Council.

Amy, D. J. (1987). Can policy analysis be ethical? In F. Fischer & J. Forester (Eds.), *Confronting values in policy analysis* (pp. 62–65). Newbury Park, CA: Sage.

Arrow, K. (1971). Economic welfare and the allocation of resources for invention. In D. M. Lamberton (Ed.), *Economics of information and knowledge* (pp. 141–159). Harmondsworth: Penguin.

Bell, D. (1973). *The coming of post-industrial society: A venture in social forecasting.* New York: Basic Books.

Bhagwati, J. N. (1988). Poverty and public policy. *World Development, 16*(5), 539–555.

Blair, M. L. (1988). VSAT systems in developing countries. In D. J. Wedemeyer & M. R. Ogden (Eds.), *Telecommunications and Pacific development: Alternatives for the next decade* (pp. 231–233). Honolulu: Pacific Telecommunications Council.

Blaug, M. (1976). The empirical status of human capital theory: A slightly jaundiced survey. *Journal of Economic Literature, 14*(3), 827–855.

Brundenius, C., & Göransson, B. (1985). *The quest for technological self reliance: The case of telecommunications in India.* Lund, Sweden: University of Lund.

Buchanan, J. M., & Tullock, G. (1962). *The calculus of consent.* Ann Arbor, MI: University of Michigan Press.

Carreon, C. C. (1976). The requirements of developing countries. *Telecommunication Journal, 43*(2), 124–133.

Casey-Stahmer, A., & Goldschmidt, D. (1985). Satellites spreading skills for rural development. *Development: Seeds of Change, 1,* 61–63.

Chenery, H., et al. (1974). *Redistribution with growth.* Baltimore: Johns Hopkins University Press.

Clippinger, J. H. (1977, September). Can communication development benefit the Third World? *Telecommunications Policy,* pp. 298–304.

Cole, K., Cameron, J., & Edwards, C. (1983). *Why economists disagree: The political economy of economics.* London: Longman.

Cole, S., & Lucas, H. (Eds.). (1982). *Models, planning, and basic needs.* New York: Pergamon Press.

Commons, J. R. (1924). *Legal foundations of capitalism.* New York: Macmillan.

Du Boff, R. B. (1983). The telegraph and the structure of markets in the United States, 1845–1890. *Research in Economic History, 8,* 253–277.

Etienne, G. (1982). *India's changing rural scene, 1963–1979.* Oxford: Oxford University Press.

Fischer, F., & Forester, J. (Eds.). (1987). *Confronting values in policy analysis: The politics of criteria.* Newbury Park, CA: Sage.

Gabel, R. (1983). Allocation of telephone exchange plant investment. In *Adjusting to regulatory pricing, and marketing realities* (pp. 452–482). (MSU public utilities papers.) East Lansing, MI: Michigan State University Press.

Goddard, J. B., & Gillespie, A. E. (1986). Advanced telecommunications and regional economic development. *The Geographical Journal, 152*(3), 383–397.

Goldschmidt, D. (1984, September). Financing telecommunications for rural development. *Telecommunications Policy,* pp. 181–203.

Goulet, D. (1979). Development as liberation: Policy lessons from case studies. *World Development, 7*(6), 555–566.

Greenberg, A. (1985). Impasse?: The US stake in Third World telecommunications development. *Journal of Communication, 35*(2), 42–49.

Griffin, K. (1981). Economic development in a changing world. *World Development, 9*(3), 221–226.

Gupta, P. P., & Ramani, S. (1981). Computer message systems for developing nations: A design exercise. In R. P. Uhlig (Ed.), *Computer message systems* (pp. 9–14). New York: North-Holland.

Hardy, A. P. (1980, December). The role of the telephone in economic development. *Telecommunications Policy,* pp. 278–286.

Hayek, F. (1944). *The road to serfdom.* London: Routledge & Kegan Paul.

Hudson, H. E. (1982). Toward a model for predicting development benefits from telecommunications investment. In M. Jussawalla & D. M. Lamberton (Eds.), *Communication economics and development* (pp. 159–189). New York: Pergamon.

Hudson, H. E. (1984). *When telephones reach the village: The role of telecommunications in rural development*. Norwood, NJ: Ablex.

Hudson, H. E. (1987). Telecommunications and the developing world. *IEEE Communications Magazine, 25*(10), 28–33.

Hudson, H. E., Goldschmidt, D., Parker, E. B., & Hardy, A. (Eds.). (1979). *The role of telecommunications in socio-economic development: A review of the literature with guidelines for further investigation*. Geneva: ITU.

Huntington, S., & Nelson, G. (1976). *No easy choice: Political participation in developing countries*. Cambridge, MA: Harvard University Press.

Information, telecommunications and development. (1986). Geneva: International Telecommunications Union.

Innis, H. A. (1956). *The fur trade in Canada: An introduction to Canadian economic history*. Toronto: University of Toronto Press.

Johnston, B. F. (1977). Food, health, and population in development. *Journal of Economic Literature, 15*(3), 879–907.

Karunaratne, N. D. (1982). Telecommunication and information in development planning strategy. In M. Jussawalla & D. M. Lamberton (Eds.), *Communication economics and development* (pp. 211–239). New York: Pergamon.

Kaul, S. N. (1988). Emerging trends in telecommunications in India, 1987–2000. In D. J. Wedemeyer & M. R. Ogden (Eds.), *Telecommunications and Pacific development: Alternatives for the next decade* (pp. 523–526). Honolulu: Pacific Telecommunications Council.

Leff, N. H. (1984). Externalities, information costs, and social benefit-cost analysis for economic development: An example from telecommunications. *Economic Development and Cultural Change, 32*(2), 255–276.

Lewis, A. (1955). *The theory of economic growth*. Homewood, IL: Richard D. Irwin.

Machlup, F. (1977). *A history of thought on economic integration*. New York: Columbia University Press.

Maitland, D. (1984). *The missing link, report of the independent commission for world-wide telecommunications development*. Geneva: International Telecommunications Union.

Mansell, R. E. (1979). *Telecommunication subsidy policy in Northwest Canada and Alaska*. Unpublished Master's thesis, Simon Fraser University, Burnaby, BC, Canada.

Mansell, R. E. (1982). The 'new dominant paradigm' in communication: Transformation versus adaptation. *Canadian Journal of Communication, 8*(3), 42–60.

Mayer, M. (1977). The telephone and the uses of time. In I.de S. Pool (Ed.), *The social impact of the telephone* (pp. 225–245). Cambridge, MA: MIT Press.

McAnany, E. G. (Ed.). (1980). *Communication in the rural Third World: The role of information in development*. New York: Praeger.

McQuail, D. (1986). Is media theory adequate to the challenge of new communication technologies? In M. Ferguson (Ed.), *New communication technologies and the public interest: Comparative perspectives on policy and research* (pp. 1–17). Beverly Hills: Sage.

Melody, W. H. (1981). The economics of information as resource and product. In D. J. Wedemeyer (Ed.), *Proceedings of the PTC* (pp. C7-5–C7-9). Honolulu: Pacific Telecommunications Council.

Melody, W. H. (1985). The information society: Implications for economic institutions and market theory. *Journal of Economic Issues, 19*(2), 523–539.

Melody, W. H. (1986). Learning from the experience of others: Lessons from social experiments in information technology in North America. *Proceedings of the Social experiments with information technology conference, Odense (Denmark), 13–15 January* (pp. 57–69). (Document no. 83, FAST Programme.) Luxembourg: Directorate-General for Science, Research, and Development, Commission of the European Communities.

Newman, G. (1976). An institutional perspective on information. *International Social Science Journal, 28,* 466–492.

O'Sullivan, J. (1980). Guatemala: Marginality and information in rural development in the Western Highlands. In E. G. McAnany (Ed.), *Communication in the rural Third World: The role of information in development* (pp. 71–106). New York: Praeger.

Oettinger, A. (1980). Information resources: knowledge and power in the 21st century. *Science, 209,* 191–198.

Parker, E. B., & Rinde, J. (1988, September). Transaction network applications with user premises earth stations. *IEEE Communications Magazine,* pp. 23–35.

Parker, E. B. (1976). Planning communication technologies and institutions for development. In S. A. Rahim & J. Middleton (Eds.), *Perspectives in communication policy and planning* (pp. 43–76). Honolulu: East-West Center.

Parker, E. B. (1981). *Economic and social benefits of the REA telephone loan program.* Mountainview, CA: Equatorial Communications.

Parker, E. B. (1982). Communication satellites for rural development. In J. R. Schement, F. Gutierrez, & M. A. Sirbu (Eds.), *Telecommunications policy handbook* (pp. 3–9). New York: Praeger.

Parker, E. B. (1984, September). Appropriate telecommunications for economic development. *Telecommunications policy,* pp. 173–180.

Parker, E. B. (1987). Micro earth station satellite networks and economic development. *Telematics and Informatics, 4*(2), 109–112.

Porat, M. (1977). *The information economy: Definition and measurement.* (Special Publication 77-12 (1).) Washington, DC: U.S. Department of Commerce, Office of Telecommunications.

Reich, R. B. (1987). *Tales of a new America.* New York: Times Books.

Samarajiva, R. A. (1985). *Property rights and information markets: Policy issues affecting news agencies and online databases.* Unpublished doctoral dissertation. Simon Fraser University, Burnaby, B.C., Canada.

Saunders, R. J. (1982). Telecommunications in developing countries: Constraints on development. In M. Jussawalla & D. M. Lamberton (Eds.), *Communication economics and development* (pp. 190–210). New York: Pergamon.

Saunders, R. J., Warford, J. J., & Wellenius, B. (1983). *Telecommunications and economic development.* Baltimore: Johns Hopkins University Press.

Schaar, J. H. (1967). Equality of opportunity, and beyond. In J. R. Pennock & J. W. Chapman (Eds.), *Equality* (pp. 228–249). New York: Atherton Press.

Scherer, F. M. (1980). *Industrial market structure and economic performance* (3rd ed.). Boston: Houghton Mifflin.

Schiller, H. I. (1976). *Communication and cultural domination.* New York: International Arts & Sciences Press.

Smith, A. (1982). *The wealth of nations.* (Reprint, with introduction by A. Skinner.) Harmondsworth: Penguin. (Original work published 1776)

Smythe, D. W. (1981). *Dependency road: Communication, capitalism, and consciousness in Canada.* Norwood, NJ: Ablex.

Streeten, P., Burki, S. J., Haq, M. U., Hicks, N., & Stewart, F. (1981). *First things first: Meeting basic human needs in developing countries.* Oxford: Oxford University Press.

Sullivan, C. (1988, May). *Overcoming the telecommunications bottleneck: The effect of information sector growth on telecommunication development policies in Ireland and South Korea.* Paper presented at the 38th conference of the International Communication Association, New Orleans, LA.

Weinhaus, C. L., & Oettinger, A. (1988). *Behind the telephone debates.* Norwood, NJ: Ablex.

Wellenius, B. (1984, March). On the role of telecommunications in development. *Telecommunications Policy,* pp. 59–66.

Zajac, E. E. (1981). Is telephone service an economic right? In H. M. Trebing (Ed.), *Energy and communications in transition* (pp. 94–109). (MSU public utilities papers.) East Lansing, MI: Michigan State University Press.

chapter 15

A Social or Economic Good? The Role of Information in the Information Society

W. Richard Goe
Senior Research Scholar
Center for Economic Development
Carnegie Mellon University

Martin Kenney
Associate Professor
Department of Applied Behavioral
Sciences
University of California, Davis

For the past several decades, social theorists and futurists have predicted that advanced industrial nations are in the nascent stages of a major social transformation. There has been a general consensus among these advocates that advanced industrial nations are undergoing an extended period of change in which the foundation for a new social structure is being prepared. In the 1970s it was predicted that the changes underway in many advanced industrial societies, especially the United States, could best be described by such labels as *postindustrial society, technetronic society, the knowledge society,* and *postaffluent society,* among others (Bell, 1973; Brzezinski, 1970; Drucker, 1969; Gappert, 1975; Touraine, 1971).[1] More recent forecasts predict that advanced industrial nations are becoming information societies—where the creation, processing, handling, and transfer of information is increasingly the most critical process underlying socioeconomic activity. Highly interrelated with this process is the growing use of technologies derived from the convergence of computers and telecommunications (Bell, 1981; Cleveland, 1982, 1985; Dizard, 1985; Masuda, 1981; Porat, 1978).

Collectively, technologies combining computers and information processing equipment with telecommunications have become known as *information technology.*[2] As this chapter will outline, information technology is contributing to a

---

[1] For a good overview of earlier forecasts, see Marien (1973).

[2] *Information* is defined as the sum total of all the facts and ideas that are available to be known by a person at any given moment in time. This is in contrast to *knowledge,* which involves refining the mass of facts and selecting and organizing what is *useful* (see Cleveland, 1982). *Technology* is defined as an organized body of knowledge which allows practical human purposes to be achieved in

broad set of changes within the U.S. political economy. Information technology has become a central element in the production of goods and services in response to global economic competition. This has created vast market opportunities for information technology-based products and services which are providing an increasingly important source of economic advantage (Cohen & Zysman, 1987; Locksley, 1986).

Broadly conceived, information technology provides: (a) the ability of computer technology to process, organize, and store information needed to coordinate or perform specific tasks (e.g., the provision of instructions for automated machinery); combined with (b) the ability to instantaneously transmit computerized information to any geographic point equipped with the necessary technology. In effect, information technology allows information to be processed, organized, stored, and transferred in order to optimally control or coordinate social activity toward desired ends. This is exemplified by the use of computer networks to trade stocks more rapidly in response to changing market conditions, or by the instantaneous transmission of corporate documents from New York to Tokyo via facsimile machine.[3] Information technology is revolutionizing the ways in which information is disseminated, exchanged, and utilized within the economy.

The purpose of this chapter is to examine the impact that information technology is having on the role of information in the U.S. political economy as part of the hypothesized transition to the information society. The first section examines specific applications of information technology and the ways in which they are transforming the use of information in the production of goods and services. The growing importance and power of information technology has created opportunities for commercial services to provide computerized information in the marketplace as an economic good. Examples include electronic databases stored on optical and compact disks and online information services such as videotex. The second section examines trends in the electronic information industry.

The advanced capabilities of information technology in facilitating the transfer and exchange of information and the development of the electronic information industry have contributed toward a heightened awareness of the role of information in the economy. The third section examines shifts in federal information policy during the 1980s. During the Reagan Administration, federal information policy actively pursued the privatization of government information

---

a reproduceable manner (Mesthene, 1970, p. 25; Brooks, 1982). The distinction has been made between *physical* technology (e.g., a machine) *social* technology, (e.g., a form of social organization), and *intellectual* technology, which consists of a set of problem-solving rules such as a computer program (see Bell, 1973, pp. 29–30).

[3] In Japan, faxed documents have been considered legally binding since the mid-1970s, thus drastically increasing the usefulness of facsimile machines.

services, involving the use of information technology and the introduction of users' fees for government documents and other forms of information.

The final section examines value considerations underlying shifts in information policy and the role of information in the economy. Information technology has the potential of providing numerous social benefits in addition to economic ones. The privatization of government electronic information services and the introduction of users' fees facilitates the establishment of electronic information as an economic good and marks a retreat from the social provision of information by the federal government. This raises the question of whether these policy strategies necessarily benefit the public interest.

THE GROWING SIGNIFICANCE OF INFORMATION TECHNOLOGY IN THE U.S. ECONOMY

Information technology is being applied across a wide range of industry sectors in the U.S. economy. In the manufacturing sector, automated manufacturing machinery such as robots, numerically controlled machine tools, and computer-aided design and computer-aided manufacturing systems (CAD/CAM) are being linked into a unified network with central computers via communications systems. This strategy has been termed *computer integrated manufacturing* (CIM) (Sillitoe, 1985, p. 82; Zygmont, 1987, p. 28).

CIM is not only allowing the reorganization of production on the factory floor, but also is allowing the reorganization of linkages within the broader firm and between firms in separate industries. There are three basic CIM strategies that are emerging. One strategy is *beginning-to-end* integration. This strategy creates an information flow that links all stages of the product development cycle from design and planning through engineering and production to support departments such as marketing and technical publications. An alternative strategy is *top-to-bottom* integration, which consists of a network of computers that links all levels of a manufacturing firm's organizational hierarchy. This type of integration strategy facilitates the flow of information from management to the production floor for better control of manufacturing operations and from the shop floor to management to enhance decision making and planning.

A final strategy is *inside-to-outside* integration, which consists of a computer network for data exchange between a manufacturer and its suppliers, distributors, and customers (Zygmont, 1987, p. 28). These strategies indicate that CIM is transforming the social linkages in the sequential system of production, distribution and support of manufactured goods. CIM allows greater economic efficiency, i.e., productivity, by providing a smoother flow of information for the more effective coordination of activities within the organizational hierarchies of firms and between firms at different stages of the production, distribution, and support system.

Through inside-to-outside integration, information technology is having a profound impact on the wholesale and retail trade sectors as well. Retail trade firms are being electronically linked with wholesale distributors in industries such as pharmaceuticals, apparel, and food distribution. For wholesale firms, computer networks allow the automation of such tasks as receiving retail product orders, dispatching orders to warehouses, and printing retail price stickers. This not only solidifies sales contracts by creating conveniences for retail firms, but also reduces labor costs by eliminating the activities of sales persons. Further, by being networked to manufacturers, wholesale distributors can provide data on product demand so production runs can be scheduled more efficiently (Hamilton & Welch, 1987). This particular form of inside to outside integration has also become known as "electronic data interchange" (Harris, Foust, & Robinson, 1987).

Within retail trade firms, point-of-sale terminals at checkout counters linked to central computers allow sales data to be instantly processed and assembled. With updated patterns of consumer demand readily available, retail sales firms can minimize costly inventories of low demand products and allocate shelf space accordingly on the basis of sales volumes (National Institute for Research Advancement, 1985, p. 51). Many large multilocational retail trade firms in the U.S. such as J.C. Penney, Wal-Mart, and K-Mart, are having their establishments linked nationwide through computer networks (see Brody, 1987, p. 43; Harris et al., 1987, p. 80).

In finance, applications of information technology such as electronic funds transfer (EFT) and automated teller machines (ATM) are transforming the banking industry. EFT allows the instantaneous transfer of funds between banking institutions, thereby reducing the time necessary for debit settlements and the exchange of money between and within banking firms (Ide, 1982, p. 71). ATMs allow bank accounts to be accessed at any time from geographically dispersed locations (Lamborghini, 1982, pp. 147–148; National Institute for Research Advancement, 1985, pp. 51–52). ATMs not only provide customer convenience, but also the possibility of reducing labor costs by replacing bank tellers (Friedrichs, 1982, p. 198).

It has been predicted that the organizational structure of banking is likely to undergo radical changes due to the application of information technology (Lamborghini, 1982, p. 147). There are signs that this may already be occurring within the United States. The banking industry in the deregulated environment of the 1980s has been characterized by waves of mergers and consolidations leading to the emergence of "super regional" banks with operations in multiple states (Guenther, 1987). The growth of interstate banking has been facilitated by information technology, since it allows greater control of geographically dispersed banking establishments.

Information technology has also had profound consequences for financial trading in stocks, securities, and other instruments. Through computerized program trading, information technology allows stocks and futures options to be traded much more rapidly (Russell, 1986). This technology has been partially blamed for the record stock market collapse in October 1987 (Ricks & Langley, 1987). In another application, electronic information services using videotex or teletext technology allow current commodities and securities quotes to be instantly obtained (Field & Harris, 1986, p. 84).[4]

Computer networking and electronic information services are also having significant impacts in a wide range of other industries in the service sector. In transportation, information technology allows electronic surveillance of shipping and trucking movements[5] and electronic booking of airline reservations.[6] In medical services, electronic information systems allow instant access to such information as medical records, patient histories, and diagnoses (National Institute for Research Advancement, 1985, p. 54). Additionally, these systems are becoming increasingly important in legal services by providing instant access to large data banks of legal information for use in writing briefs or other legal procedures.[7]

In administrative and management functions all industry sectors including government are automating their offices. Computers of all sizes, and office work stations, are being linked together in networks to allow information to be more easily exchanged and shared among office workers (Wilson & Schiller, 1987, p. 112). Other technologies being utilized in this process are electronic mail, voice mail, facsimile machines, and videoconferencing. These applications are facilitating the more efficient exchange of information within firms (whether multilocational or single establishments) and between separate businesses. The growing use of these technologies may lead to the "paperless" office, where all data,

---

[4] For example, as an independent company, prior to its merger with Shearson-Lehman Brothers, E. F. Hutton utilized a telecommunications and satellite network to deliver financial information to its 400 retail offices (Field & Harris, 1986, p. 84). This technology has become essential to security and commodity brokerage firms and is also increasingly being used by investors.

[5] Mardata, owned by Lloyd's Maritime Data Network, Inc., and Goestar, owned by Geostar, Inc. are examples of shipping and trucking services (see Field & Harris, 1986, p. 84; Wessel, 1986). Through using satellite networks, these services allow the global location of individual ships and trucks to be instantly pinpointed. Other applications of information technology allow the location and status of goods being transported to be continuously tracked and monitored. One example of this type of application is Federal Express Corporation's COSMOS IIB system.

[6] These systems have become a source of controversy in the airline industry. The five systems that are operative are owned by the large carriers (United Airlines, American Airlines, Texas Air, Transworld Airlines, and Delta Airlines). Other airlines have charged that these systems have given the large carriers an unfair competitive advantage, since they must pay fees to have their flights listed, and close to 70% of all airline tickets are sold by travel agents who utilize these systems. Additionally, it is charged that these systems are architecturally biased as the owning company's flights are listed first which promotes a disproportionate amount of sales to these carriers (McGinley, 1987).

[7] An example is Mead Corporation's LEXIS electronic information service, which is currently used by most major law firms (Schiller, 1986, p. 90).

documents, and diagrams are written, processed, stored, and exchanged electronically (National Institute of Research Advancement, 1985, p. 51).

In all of these cases, information technology is electronically providing instantaneous access to computerized information across geographic space. This makes it possible to improve the coordination of planning, decision making, administration and production activities. This capability has made some individuals and organizations willing to pay for access to computerized information, and an industry is taking shape in the United States.

TRENDS IN THE ELECTRONIC INFORMATION INDUSTRY

Information technology has provided new means of transforming information into a proprietary economic good that extend beyond previously developed mass media. Online, computerized data services (e.g., videotex), and compact and optical disk technology provide new means of selling information as an economic good. The development of compact and optical disk technology for use in storing computerized information has occurred only recently. As information storage media, these technologies have a much greater storage capacity, compared to conventional floppy disks.[8] A number of firms have begun developing and marketing large data bases stored on compact and optical disks.[9] However, these data bases are generally high in cost, and the peripheral hardware necessary to read and/or write information on compact and optical disks has yet to become widely adopted and utilized (Bulkeley, 1985, 1986; Miller, 1986; *Wall Street Journal*, 1986). It remains to be seen how pervasive the use of this laser-based technology will become, compared to conventional electromagnetic storage media.

A movement toward marketing computerized information via online services gained momentum during the late 1970s. Many of these services have been unsuccessful in developing adequate markets. This has been especially true of videotex services, which have marketed other services such as electronic banking and home shopping in addition to retrieval of computerized information (Connelly, 1984, 1985; Field & Harris, 1986; Harris, 1985). Several ventures, however, have managed to successfully develop mass markets for electronic information. In 1985, revenues of the top eight electronic information firms ranged from $100 to $505 million. The market was dominated by the sale of

---

[8] For example, a 14″ optical disk developed by Kodak has a storage capacity equivalent to 2,000 floppy disks (*Wall Street Journal*, 1986).

[9] A number of large financial and business data bases have been developed ranging from historical prices of stocks and bonds to information on U.S. corporations. Other products include sets of encyclopedias and databases on aerospace engineering and environmental health and safety (Bulkeley, 1985, 1986).

business and financial information, such as securities and commodities quotes and credit information (Field & Harris, 1986, p. 84).[10]

It was initially believed that a mass market for electronic information could be quickly developed among home consumers. Videotex was originally conceived as a technology that could deliver information and electronic newspapers, provide teleshopping, telebanking services, and the booking of airline reservations—all within the home. However, the home consumer market for videotex has remained small. Only a few videotex ventures, including The Source and CompuServe, have managed to establish adequate markets. There has been a reluctance among home consumers to pay for the high cost of electronic information and additional services they are not sure they need (Harris, 1985).

Businesses have become the primary consumers of electronic information. Firms in industries with information intensive needs (e.g., finance, legal services) have recognized the value of having instant access to information provided by electronic information services. In cases where firms cannot internally generate and transfer such information to workers in an efficient manner, electronic information services provide an alternative. In this respect, electronic information is becoming a purchased input used in the production of goods and services.

As a commodity, information has characteristics not possessed by other, more tangible products. When consumed, a supply of information is not depleted.[11] When the possessor of information either sells or shares it with others, control over the information is not surrendered and it is then possessed by all parties who have acquired it; i.e., it becomes a collective good. Further, the shelf life of different types of information is variable. Some information retains its value indefinitely, while other information is obsolete seconds after it has been generated (Bell, 1981, p. 512; Compaine, 1981, p. 133; Jones, 1985, p. 225; Pool, 1981, 158–159). These attributes make it extremely difficult to gain the degree of proprietary control over information needed to capture the full economic benefits from its sale in the marketplace. Due to the ability to share information as a collective good without distributing it via market relations, proprietary control can be most attained over *initial* access to information.

As previously outlined, one important dimension of the hypothesized transition to the information society is the growing role of information technology in generating, transferring, and processing information for use in the production of goods and services. A second important dimension is that access to electronic

---

[10] The major firms in the electronic information industry in 1985 included Reuters, Dun & Bradstreet, Quotron, TRW, Mead, Telerate, McGraw-Hill, and Dow Jones (see Field & Harris, 1986, p. 84).

[11] *Consumption of information* refers to the processing of information by the human mind, a computer, or computer-based machine.

information that is not internally generated within a particular firm, organization, or home will likely have to be purchased in the marketplace, whether in the form of a commercial online service or through a product such as a database stored on compact disk. While this has long been the case with previously established mass media such as newspapers, magazines, and cable television, information technology further extends the "commodification" of information.

The movement to establish electronic information as an economic good has not been restricted to the efforts of the private sector to create commercial products and services. The federal government has actively supported this movement through the privatization of government information services involving the use of information technology. This has been the result of shifts in federal information policy that have occurred over the past decade.

SHIFTS IN FEDERAL INFORMATION POLICY AND THE MOVEMENT TOWARD THE PRIVATIZATION OF PUBLIC DOMAIN INFORMATION

The movement to reorganize the federal government's mechanisms for disseminating information was initiated during the Carter Administration. This was a result of several forces. First, there was the decision that the number of federal agencies disseminating public information had grown too large in number, resulting in unnecessary duplication and overlap in information being collected and disseminated (Office of Management and Budget, 1978, 32,204–32,305). Second, since World War II, there has been a constant increase in the linkages between the Government Printing Office and private sector printing firms which have handled an increasingly larger percentage of government printing contracts. Through the Information Industry Association, the private sector constantly pressured the GPO to become the main distributional channel of government-generated information. The principal argument of the Information Industry Association was that government information dissemination should be handled privately and contribute to profit making (Schiller, 1981, pp. 59–62).

The consideration of information as an economic resource was embraced and supported by agencies within the federal government. For example, a 1980 statement issued by the National Telecommunications and Information Administration (a branch of the U.S. Department of Commerce) entitled "The Foundations of United States Information Policy" clearly reflects this position in describing the properties of information:

> Information is a resource. Like energy, capital, or labor, information is a resource that can be applied to achieve economic, social, or political goals.

> Information is a commodity. Information is sold, traded, or otherwise exchanged, frequently for financial or other reward to the person or organization sharing or giving up the information. (Bushkin & Yurow, 1980, p. 3)

The drive to privatize government-generated information resulted in the drafting of the Federal Publications Act of 1980, which included: (a) a provision for the reorganization of all government information activities under a single entity, entitled the Federal Publications Office (FPO), that would be controlled by the legislative branch; (b) a provision that production and distribution of information materials may be contracted with the private sector if the service cannot be satisfied by the FPO, or if the contract will result in cost savings to the government; and (c) a provision that all government publications be sold at a price which shall not be less than the cost of production and distribution (United States Congress, 1980). Although the act was not passed in the 96th Congress, the policy concepts it articulated were influential in shaping subsequent federal policy initiatives.

Immediately following the ill-fated Federal Publications Act came the Paperwork Reduction Act of 1980, which reflected the earlier efforts to reorganize federal information activities. The primary ostensible purpose of the Paperwork Reduction Act was "to minimize the federal paperwork burden for individuals, small businesses, state and local governments and other persons" (United States Statutes At Large, 1980, 84 STAT. 2812). However, there were several other important provisions of the bill which authorized significant reorganization of federal information activities.

One important provision of the bill was the establishment of an Office of Information and Regulatory Affairs within the Office of Management and Budget (OMB). The director of OMB was vested with the authority to "develop and implement federal information policies, standards, and guidelines and shall provide direction and oversee the review and approval of information collection requests, the reduction of the paperwork burden, federal statistical activities, records management activities, privacy of records, interagency sharing of information and acquisition and use of automatic data processing telecommunications, and other technology for managing information resources" (United States Statutes At Large, 1980, 94 STAT. 2815).

The mandated activities to be carried out by the director of OMB included: (a) a 25% reduction in the amount of information collected by the federal government over the next 2 years; and (b) the identification of areas of duplication in information collection requests and the development of a schedule and methods for eliminating duplication. Directors of all federal agencies were ordered to review and conduct their own information management activities in an efficient, economical manner, ensure that their information systems did not overlap or duplicate those of other agencies, and comply with the information policies, principles, standards, and guidelines prescribed by the director of OMB (United States Statutes At Large, 1980, 94 STAT. 2818-94, STAT. 2819).

Another important provision of the Paperwork Reduction Act was "to ensure that automatic data processing and telecommunications technologies are acquired and used by the federal government in a manner which improves service delivery and program management, increases productivity, reduces waste and

fraud, and, wherever practicable and appropriate, reduces the information processing burden for the federal government and for persons who provide information to the federal government" (United States Statutes At Large, 1980, 95
STAT. 2812-94, STAT. 2813). This created a coordinating mechanism for the
incorporation of information technology into the information management and
dissemination activities of the federal government.

The Paperwork Reduction Act was signed into law (PL96-511) by President
Carter on December 11, 1980. In essence, it provided OMB (a key agency
within the executive branch) with the sole authority to set federal information
policy based on an extremely broad set of guidelines. This allowed the Reagan
Administration to tailor federal information policy to its conservative agenda.
The heavy emphasis of the Reagan Administration on paring government programs, and privatizing and recovering the cost of government services, was
influential in shaping federal information policy during the 1980s. This is summarized in OMB Circular A-130 entitled "Management of Federal Information
Resources," which was issued on December 12, 1985.

One of the underlying assumptions of this policy statement is the recognition
of information as a proprietary economic good.

> Government information is a valuable national resource. It provides citizens with
> knowledge of their government, society, and economy—past, present, and future;
> is a means to ensure the accountability of government; is vital to the healthy
> performance of the economy; is an essential tool for managing the government's
> operations; *and is itself a commodity often with economic value in the marketplace.*
> (OMB, 1985, p. 3; emphasis added)

A second basic assumption is that the federal government should rely primarily
on the private sector to provide government services.

> Although certain functions are inherently governmental in nature, being so inti-
> mately related to the public interest as to mandate performance by federal em-
> ployees, *The government should look first to private sources,* where available, to
> provide the commercial goods and services needed by the government to act on
> the public's behalf, particularly when cost comparisons indicate that private perfor-
> mance will be the most economical. (OMB, 1985, pp. 3–4; emphasis added)

As part of the overall information policy, federal agencies were required to:
(a) place maximum feasible reliance on the private sector for the dissemination
of information products and services, and (b) recover the costs of disseminating
the information products or services through user charges where appropriate
(OMB, 1985, p. 5). Regarding the use of information technology, federal agencies are mandated to meet information processing needs through interagency
sharing and from commercial sources, when it is cost effective, before acquiring
new information processing capacity (OMB, 1985, p. 6).

This trajectory in information policy has resulted in a large volume of information that was previously disseminated through "in-house" government channels now being rechanneled through the private sector. The goal of reducing government information collection and dissimination has been carried out, and user fees have been applied to many government publications that were previously distributed free of charge (Demac, 1984, pp. 32–34). While the Paperwork Reduction Act accelerated the use of information technology in the provision of government information services, it has also resulted in the privatization of government information services. Many computerized databases developed by the federal government are now being distributed by private sector electronic information services. These changes are illustrated in the policies employed by the USDA in its system for electronically distributing agricultural information.

In the late 1970s, the USDA began creating online databases and also subsidized experiments with agricultural videotex and teletext systems (Goe, 1988, pp. 130–131; Lett, 1983, p. 127; United States Congress, 1983). The privatization of USDA electronic information services gained momentum with the establishment of an electronic mail network that linked USDA and Cooperative Extension Service offices. In 1981 the provision of this service was contracted to the data-processing firm Dailcom, Inc., which was also serving other federal agencies. In 1982 the USDA expanded the electronic mail service to include access to several of its online databases. The expanded service was entitled USDA ONLINE. In 1982 Dailcom was acquired by ITT, which began to provide the USDA ONLINE service.

In 1984 the USDA decided to create an electronic information service that would be the sole repository for all USDA "perishable" information. Types of information that are defined as perishable include USDA market reports, crop and livestock statistical reports, economic outlook and situation reports, foreign agricultural trade leads, export sales reports, world agricultural roundups, and USDA press releases (Martin Marietta Data Systems, 1985, p. 1). The provision of this service was contracted to Martin Marietta Data Systems. The service was labeled the Electronic Dissemination of Information System (EDI) and went online in 1985.

The contractual arrangements between the USDA and ITT and Martin Marietta Data Systems are similar in nature. The USDA subsidizes the cost of data entry and storage. ITT and Martin Marietta Data Systems then distribute the information for the cost of accessing their services and technically are not allowed to charge for the information (although Martin Marietta has a per line change). In effect, the primary consequence of the privatization of USDA electronic information dissemination activities is that public domain information must be obtained through commercial services with access being sold in the marketplace. Information that was once distributed as a free public service must now be purchased.

While privatization may be more economically efficient for the federal government in a time of budgetary deficits, it increases the cost of information for the consumer. The cost of private sector dissemination (labor, technology, etc.) must be recovered, whether in the price of the information or through government subsidies. As it becomes more widespread, the establishment of electronic information as an economic good raises a number of value considerations and policy issues.

VALUE CONSIDERATIONS AND POLICY ISSUES UNDERLYING THE ESTABLISHMENT OF ELECTRONIC INFORMATION AS AN ECONOMIC GOOD

The privatization of government information services reflects a marked shift in values regarding the role of public domain information in the economy. Prior to the Paperwork Reduction Act, federal information policy assumed that the public interest could best be served by providing a source of information, independent of private interests, that could be accessed by all at no charge. Current information policy regarding electronic dissemination assumes that the public interest can best be served by accommodating the interests of the private sector, regardless of the impact this may have on the public's ability to access public domain information.

Persons desiring to access public domain information via information technology must pay the high cost of private electronic information services. Additionally, members of the public wishing to obtain paper documents and other forms of information distributed "in-house" by the federal government must, in most cases, pay users' fees. The privatization of electronic information services, and the employment of users' fees, extends the concept of information as an economic good and marks the retreat of the federal government from the social provision of information as a free public service.

This raises a number of value considerations, since government information is generated from public tax revenues. Should the U.S. public be forced to pay for the generation of the information and then a second time for access to it? Like information services produced solely by the private sector, the introduction of users' fees and the privatization of government electronic information services have the potential of limiting access to those with the ability to pay. It is currently unknown what effects privatization and users' fees have had on the public's ability to access public domain information.

The drive to establish electronic information as an economic good has the potential of limiting the social benefits of information technology. With the ongoing process of creating and enlarging computerized data bases, information technology places a much greater array of information resources at the fingertips of those possessing the technical skills to access it. This can allow a greater number of individuals, organizations, and businesses access to information needed to optimally coordinate their activities toward desired ends. If available to all, this capability could benefit the broader public interest as well as the narrower-range interests of specific persons, organizations, and business firms.

Access to computerized information networks will be determined by information policy. Whether access is democratically distributed or is restricted to only a portion of society represents a public choice that will influence the distribution of the benefits of the information society. In contrast to the U.S., Japan is one advanced industrial nation that appears to be favoring the course of democratic participation. The utopian plan of the information society developed by Japan's Institute for the Information Society emphasizes that *all* Japanese households must ultimately have access to computerized information networks in order to obtain information to resolve problems and to pursue new possibilities of the future. These include training and self-development through computerized education programs and a wider range of opportunities for choice when selecting future work or the direction of social activity (Masuda, 1981, pp. 39, 65, 66).[12]

The Japanese plan stresses that computerized data bases (entitled information utilities) must ultimately be controlled and managed by Japanese citizens. Only this can fully ensure that the privacy of personal data (e.g., tax records) is maintained while the benefits of public information resources are made available to all (Masuda, 1981, p. 111; National Institute for Research Advancement, 1985, p. 92). In essence, the Japanese plan stresses the long-term importance of employing information technology as a tool for social development through informing and educating the Japanese public while protecting the rights of citizens.

In contrast, the current trajectory of U.S. information policy appears to disregard this possibility. Information policy under the Reagan Administration emphasized the short-term commercial potential of electronic information services over any long-term social benefits of developing public access electronic information networks. The social benefits of electronic information networks were ignored in the rush to capture the economic benefits of information technology and ease the massive federal budget deficit. Faced with this problem, no substantial change has occurred under the Bush Administration.

The social provision of information by the federal government has historically provided an important function in the development of U.S. society. For example, the free dissemination of information on new agricultural production techniques by the Cooperative Extension Service played a key role in the development of U.S. agriculture in the twentieth century. Further, the wide range of

---

[12] This implicitly assumes that labor processes and social exchanges will be increasingly mediated by information technology. Thus, telecommuting will likely become more prevalent and a larger number of social interactions will take place through such forms of information technology as electronic mail, voice mail, and teleconferencing.

information services provided through the federal repository library system and numerous federal agencies has played an important role in informing the functions of researchers, educators, activists seeking to resolve social problems, and businesses alike. Privatization and the employment of users' fees have the potential of restricting access to government information and limiting its accompanying social benefits.

The social benefits foregone must be incorporated into the calculus of decision making regarding the further extension of privatization and the transformation of government information into an economic good. Privatization should not be deemed a successful strategy to be continued or extended unless it is found that this strategy has had little or no effect in prohibiting access to public domain information by U.S. taxpayers. Similarly, users' fees should also be rescinded if found to be inhibiting access. Whether a return to the social provision of information is in the best public interest needs to be carefully evaluated. This is especially vital given the growing importance of information technology and the need to access electronic information.

An important related issue is whether a free flow of electronic information is desirable. The Reagan Administration actively encouraged electronic information firms and libraries to police their customer bases for users who might be potentially unfriendly to the United States. The federal government has also initiated a movement aimed at restricting access by such users to computerized databases containing "sensitive" information in the interest of national security (Davis, 1987; Starr, Port, Schiller, & Clark, 1986). In addition to public sector initiatives, efforts to curb access to electronic information have also been undertaken by the private sector. For example, Dun & Bradstreet Corporation has blocked access to its financial data bases by labor unions, claiming this represents unauthorized use of its data (Roberts, 1987). In effect, restricting the free flow of electronic information results in discriminatory access, even among those who possess the financial resources to purchase access. This has the potential of violating citizen's rights and exacerbating inequity in access.

The movement to establish electronic information as an economic good in the United States involves several aspects that potentially threaten to restrict the social benefits of electronic information services. As the creation, transfer, and exchange of information via information technology becomes more prevalent in guiding socioeconomic activity, the extent of access to electronic information will play an influential role in shaping societal development. If the United States is undergoing a transformation to an information society, an "information bill of rights" is needed that will ensure the rights of access to electronic information and maintain the privacy of personal data. An optimal information policy would be one that strikes a balance between promoting the broader range social benefits of information technology and the narrower range economic benefits subsumed beneath them.

REFERENCES

Bell, D. (1973). *The coming of post-industrial society.* New York: Basic Books.

Bell, D. (1981). The social framework of the information society. In T. Forest (Ed.), *The microelectronics revolution* (pp. 500–549). Cambridge, MA: The MIT Press.

Brody, H. (1987). Big hopes for small dishes. *High Technology Business, 7*(11), 41–45.

Brooks, H. (1982). Social and technological innovation. In S. B. Lundstedt & E. W. Colgalazier (Eds.), *Managing innovation* (pp. 1–30). New York: Pergamon Press.

Brzezinski, Z. (1970). *Between two ages: America's role in the technetronic era.* New York: The Viking Press.

Bulkeley, W. M. (1985, November 4). Compact disk technology is finding new niches in computer data field. *The Wall Street Journal,* p. 16.

Bulkeley, W. M. (1986, September 23). Lotus plans pioneering step in marketing financial information on compact disks. *The Wall Street Journal,* p. 4.

Bushkin, A. A., & Yurow, J. H. (1980). *The foundation of United States information policy.* Washington, DC: U. S. Department of Commerce.

Cleveland, H. (1982). Information as a resource. *The Futurist, 16*(6), 34–39.

Cleveland, H. (1985). The twilight of hierarchy: Speculations on the global information society. *Public Administration Review, 45,* 185–195.

Cohen, S. S., & Zysman, J. (1987). *Manufacturing matters: The myth of the post-industrial economy.* New York: Basic Books.

Compaine, B. M. (1981). Shifting boundaries in the information market-place. *Journal of Communication, 31* (1), 132–142.

Connelly, M. (1984, November 2). Knight-Ridder's cutback at Viewtron Shows videotex revolution is faltering. *The Wall Street Journal,* p. 43.

Connelly, M. (1985, March 28). Ailing Videotex ventures haven't slowed plans to market the information service. *The Wall Street Journal,* p. 33.

Davis, B. (1987, September 23). Federal agencies press data-base firms to curb access to 'sensitive' information. *The Wall Street Journal,* p. 4.

Demac, D. A. (1984). *Keeping America Uninformed. Government secrecy in the 1980s.* New York: The Pilgrim Press.

Dizard, W. P. (1985). *The coming information age: An overview of technology, economics and politics* (2nd ed.). New York: Longman.

Drucker, P. F. (1969). *The age of discontinuity.* New York: Harper & Row.

Field, A. R., & Harris, C. L. (1986, August 25). The information business: Despite a slow start, many still see a bonanza in selling electronic data. *Business Week,* pp. 82–90.

Friedrichs, G. (1982). Microelectronics and macroeconomics. In G. Friedrichs & A. Schaff (Eds.), *Microelectronics and society: For better or for worse* (pp. 189–211). New York: Pergamon Press.

Gappert, G. (1975). *Post-affluent America: The social economy of the future.* New York: New Viewpoints.

Goe, W. R. (1988). *Food production in the emerging information society: A political-economic analysis.* Unpublished doctoral dissertation, The Ohio State University, Columbus, OH.

Guenther, R. (1987, October 1). Some regional banks grow rapidly, reach major league status. *The Wall Street Journal,* pp. 1, 18.

Hamilton, J., & Welch, R. (1987, December 21). How Levi Strauss is getting the lead out of its pipeline. *Business Week,* p. 92.

Harris, C. L. (1985, January 14). For Videotex, the big time is still a long way off. *Business Week,* pp. 128, 132–133.

Harris, C. L., Foust, D., & Robinson, M. (1987, August 3). An electronic pipeline that's changing the way America does business. *Business Week,* pp. 80–82.

Ide, T. R. (1982). The technology. In G. Friedrichs & A. Schaff (Eds.), *Microelectronics and society: For better or for worse* (pp. 37–88). New York: Pergamon Press.

Jones, M. G. (1985). Deindustrialization of the American economy: The impact of antitrust law and policy. In R. D. Buzzell (Ed.), *Marketing in an electronic age* (pp. 223–237). Boston: Harvard Business School Press.

Lamborghini, B. (1982). The impact on the enterprise. In G. Friedrichs & A. Schaff (Eds.), *Microelectronics and society: For better or for worse* (pp. 119–156). New York: Pergamon Press.

Lett, R. D. (1983). Computers: The newest technology for American farmers. *The Information Society, 2* (2), 121–129.

Locksley, G. (1986, Winter). Information technology and capitalist development. *Capital & Class, 27,* 81–105.

Marien, M. (1973). Daniel Bell and the end of normal science. *The Futurist, 7,* 262–269.

Martin Marietta Data Systems. (1985). *Electronic dissemination of information system: System features and data.* Green Belt, MD: Author.

Masuda, Y. (1981). *The information society as post-industrial society,* Tokyo: Institute for the Information Society.

McGinley, L. (1987, February 3). U.S. probes airline reservation systems over complaints they curb competition. *The Wall Street Journal,* p. 14.

Mesthene, E. G. (1970). *Technological change.* Cambridge, MA: Harvard University Press.

Miller, M. W. (1986, June 3). Apple enters pact to study new uses of compact disks. *The Wall Street Journal,* p. 18.

National Institute for Research Advancement. (1985). *Comprehensive study of microelectronics 1985.* Tokyo: Author.

Office of Management and Budget. (1978). Dissemination of technical information. Request for comment. *Federal Register, 43* (143), 32204–32205.

Office of Management and Budget. (1985). *Management of federal information resources.* Circular No. A-130, 12 December.

Pool, I. D. (1981). International aspects of telecommunications policy. In M. L. Moss (Ed.), *Telecommunications and productivity* (pp. 156–167). Reading, MA: Addison-Wesley.

Porat, M. U. (1978). Communication policy in an information society. In G. O. Robinson (Ed.), *Communications for tomorrow: Policy perspectives for the 1980s* (pp. 3–60). New York: Praeger Publishers.

Ricks, T. E., & Langley, M. (1987, October 21). Congress puts on fast track regulations for Wall Street. *The Wall Street Journal,* p. 21.

Roberts, J. L. (1987, December 7). Dun & Bradstreet blocks data access to labor unions. *The Wall Street Journal,* p. 10.

Russell, G. (1986). Manic market: Is computer-driven stock trading good for America? *Time, 128* (19), 64–70.

Schiller, H. I. (1981). *Who knows: Information in the age of the Fortune 500.* Norwood, NJ: Ablex Publishing Corp.

Schiller, Z. (1986, August 25). Mead makes information pay—most of the time. *Business Week,* p. 90.

Sillitoe, P. (1985, October 31). Get smart—that's the message for future factories. *Far Eastern Economic Review,* pp. 82–84.

Starr, B., Port, O., Schiller, Z., & Clark, E. (1986, December 1). Are data bases a threat to national security? *Business Week,* p. 39.

Touraine, A. (1971). *The post-industrial society.* New York: Random House.

United States Congress. (1980). *Federal Publications Act of 1980.* Report by the Committee on Government Operations of the U.S. House of Representatives, 96th Congress, 2nd session, House Report No. 96-836, Part 3. Washington, DC: U.S. Government Printing Office.

United States Congress. (1983). *Information technology for agricultural America.* Washington, DC: U.S. Government Printing Office.

United States Statutes At Large. (1980). Public Law 96-511, *United States Statutes at Large,* 96th Congress, 2nd Session, 1980, Volume 94, Part 3. Washington, DC: U.S. Government Printing Office.

Wall Street Journal. (1986, November 7). Kodak to swap data on the development of big storage disks. *The Wall Street Journal,* p. 18.

Wessel, D. (1986, March 21). Satellites add more Accuracy to Locating Objects on Earth. *The Wall Street Journal,* p. 29.

Wilson, J. W., & Schiller, Z. (1987, April 27). The nerve-racking job of setting up a computer network. *Business Week,* pp. 112–114.

Zygmont, J. (1987). Manufacturers move toward computer integration. *High Technology Business,* 7 (2), 28–33.

V
Normative Issues

chapter 16
Value and Policy Issues in the Marketplace for Broadcast Licenses

Joseph M. Foley
Professor of Communication
Chair
Department of Communication
The Ohio State University

The F.C.C. has created a lively marketplace for the transfer of broadcast station licenses. The regulations which guide this marketplace result in a situation in which the present broadcast licensees are free to transfer their licenses to anyone who is minimally qualified to be a licensee.

This chapter examines the general parameters of this marketplace, the evolution of the current regulatory policies, and the value arguments which underlie the establishment of these policies.

Any policy which is adopted in this area requires a balancing of the rights of the public to have the electromagnetic spectrum used in the ways that will most benefit the "public interest, convenience, and necessity."

Much of the debate hinges on the disagreements over whether government regulation on an unrestricted marketplace with the potential for immense profits will result in the most desirable use of the spectrum.

Examining the dynamics of this marketplace provides an opportunity to see the implications of the potential confusion between the two meanings of value. Often the term *value* is used to characterize the broad societal goals which broadcast stations can serve. Here the concern is for *social values*. Increasingly the term *value* is being used to describe the financial worth of the station. In these uses the emphasis is on *economic values*. The fundamental issue is the impact of policies which seek to optimize the economic worth of each individual station on the overall social value goals for the broadcast system.

PARAMETERS OF THE CURRENT MARKETPLACE—PRICE AS THE MEASURE OF VALUE

The marketplace in broadcast station licenses has been very active for the past few years. In addition to the sales of individual stations, there have been many group sales. Even the networks have found themselves becoming the targets of other corporations. ABC was purchased by Capital Cities Communications. NBC was acquired by General Electric, and CBS had to work to fight off a series of takeover attempts.

The market for stations has shown substantial growth over the past several decades both in the number of stations transferred each year and in the dollar volume of the transfers ("34 Years", 1988).

The left scale in Figure 16.1 shows the growth in the annual sales of radio stations from $10 million in 1954 to $1.2 billion in 1987. Even with this growth represented on a logarithmic scale, the rate of increase is dramatic. The right scale in Figure 16.1 shows the number of radio stations transferred each year has ranged from 187 in 1954 to a high of 1,558 in 1986.

The transfer market growth has been equally dramatic for television stations.

The left scale in Figure 16.2 shows that the sales of television stations ranged from an annual low of $15 million in 1959 to a high of $3.3 billion in 1985. The right scale in Figure 16.2 indicates that the number of television stations transferred annually ranged from 16 in 1962 to 128 in 1986.

Although no station is "typical," averaging a number of sales over a period of time gives a sense of the increases in the prices paid for station transfers. This is shown in Table 16.1. With such a of rapid increase in prices, it is understandable that many broadcasters have decided to sell their stations to new operators. Not

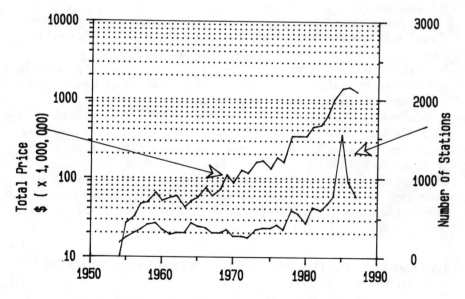

Figure 16.1.   Radio Station Transfers 1954–1987

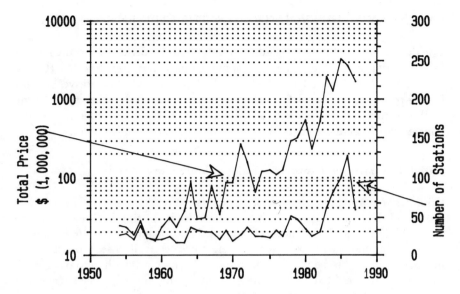

Figure 16.2.   Television Station Transfers 1954–1987

surprisingly, the new licensees who have paid these growing prices have a strong interest in maintaining the value of their property for future sale. The new licensees all hope these growth curves will continue for some time to come. This gives rise to a significant concern that the enterprises which have been attracted to broadcasting in the 1980s may be more interested in the potential for profits from the sale of their stations than they are in providing programming to serve the needs and interests of their communities.

These prices are far above the replacement costs of the physical facilities, staff, and program rights which are being purchased. The bulk of the purchase price is paid to obtain the license to broadcast.

Table 16.1.   Average Station Transfer Prices.

Time Period	Radio	Television
1954–59	$ 120,000	$ 2,340,000
1960–69	189,000	3,610,000
1970–79	480,000	7,040,000
1980–87	1,010,000	14,630,000

VALUES AND THE EVOLUTION OF THE CURRENT POLICIES

The development of policy regarding transfers can be divided into three areas: (a) an initial de facto policy which approved most proposed transfer, (b) the brief "Avco Rule" period in which the F.C.C. allowed competitive bids for license transfers, and (c) a time of increasing reliance on the marketplace. One of the features of this marketplace orientation has been a reduction in the policies which were designed to promote diversification of licensees.

Initial Treatment of Transfers—Valuing the Public Interest

From the beginning of Federal Radio Commission regulation in 1927 the policy was established that license transfers requests from existing licensees would be approved if the proposed license met the basic qualifications for holding a license. The primary criterion used to select by present licensees to select new licensees appeared to be the amount the new licensee was willing to pay for the station. If this resulted in a profit (or a loss) the entire proceeds from the transaction went to the present licensees.

Throughout this period the Commission appears to have felt increasing frustration with its inability to use higher standards to evaluate prospective transferees. In its procedures for the granting of new frequencies and for the renewal of licenses, the Commission developed a series of explicit measures to determine whether service met the values embodied in the "public interest, convenience, and necessity" standard. With license transfers, it only could use the more limited standard of basic qualifications.

The possibility that licenses could be sold to the highest bidder was a source of concern from the earliest days of broadcasting. That concern appears to have been given its first formal expression in a resolution adopted by the Fourth Annual Radio Conference in 1925:

> Resolved, that this conference views with considerable apprehension and disfavor any practice contemplating the sale of a wave length, and that we earnestly recommend that all future propositions of this kind be scrutinized most carefully by the Department of Commerce, so as to eliminate the possibility of speculating in wave lengths. (Proceedings, 1925)

The basis for this concern was the fear that broadcasters would operate their stations in ways that would enhance their "speculative" worth, rather than operate their stations in ways that would best serve the public. It was feared that this would attract other speculators to the industry who would be less concerned with serving the public than they were with obtaining profits from the sale of their stations.

There are many examples of substantial profits made from the sale of broad-cast stations. Even during World War II station sale prices remained high.

During 1944 thirty-two radio stations sold for a total of more than ten million dollars. Most of them were small stations. Most of them sold at prices which represented fantastically high profits—so high, indeed that the Federal Commu-nications Commission got worried and asked Congress what to do about it.

Here are some typical cases:
Station WINX in Washington, a little five-year old 250-watt station the net value of whose assets at the time was (according to records filed with the F.C.C.) about $75,000 sold for $500,000.

WCOP, Boston, a 500-watt station which was bought in 1936, sold for $225,000. KECA, Los Angeles, a 5,000-watt station the book value of whose tangible broad-cast property was no more than 70,000, went for $800,000. (Smith, 1945, p. 315)

Many people believed that prices substantially higher than the worth of the tangible property indicated a serious problem which required attention. At its root this problem was a fear that the industry was operating on an incorrect set of values. The prospects for these short-term profits were seen to be at odds with the long-term goal of attracting broadcasters who would serve the "public in-terest, convenience, and necessity."

The key to a successful policy required balancing the rights of the broad-casters to earn a reasonable profit with the rights of the public to receive substantial service. The focus of concern was whether the prices were so high that they precluded substantial service.

Before the Commission can properly evaluate the factors justifying a sale price, it must have a yardstick for measuring the appropriate value. On the one hand, uncontrolled prices tend to restrict prospective transferees to persons and concerns of substantial wealth and this is undesirable. On the other hand, to limit the sales price of a station to the value of its physical facilities is to deprive a licensee, who has built up a good station, of the just reward for his enterprise. If upon retiring from business a licensee was not entitled to reasonable compensation for the good will he had built up, the tendency might be for licensees to attempt to reap the fullest possible profit from their day-to-day operations at the expense of develop-ment of their station on a sound basis devoted to principles of public service. (Crosley, 1945, pp. 22–23)

Several value questions became important in the debate. Should broadcast station licenses be available to those who did not have "substantial wealth?" What is "reasonable compensation" for a licensee who wishes to leave broad-casting? Should station prices be limited so that new licensees did not have to resort to excessive commercialization to generate the revenues needed to pay for the license? What is the proper role for the F.C.C. to play in the selection of transferees? Does the F.C.C. have responsibility to ascertain that the license is transferred to the best qualified new licensee? These concerns resulted in the promulgation of the Avco rule.

The Avco Rule Period—Trying to Select the Best Future Licensee

The concerns indicated above caused the F.C.C. to attempt to take a more active role in the transfer of licenses. The precipitating incident for this action was the Commission's dissatisfaction with the existing policies which forced it to approve the transfer of WLW in Cincinnati, Ohio, from Crosley Corp. to The Aviation Corp.

The commission majority decided to approve the transfer, even though it agreed that Avco was not as well qualified as Crosley.

> If we were to judge the relative abilities of Powel Crosley, Jr. and the Aviation Corporation, clearly the transferor's long experience would outweigh any merits of the transferee's proposal. (Crosley, 1945, p. 21)

Since the majority was going to vote to approve the transfer, it did not go on to detail the shortcomings of Avco.

The dissent of Commissioners Walker and Dun presented the objections more clearly. Crosley was an established and respected broadcaster. Avco was seen as an upstart conglomerate with no commitment to the values traditionally espoused for broadcasting.

> The transferee, Aviation Corp., is a holding company. It and those now in control of its policies are engaged in activities ranging from the manufacture of kitchen sinks to the conduct of a stock brokerage business, including the manufacture of airplanes, ships, steel, and the control of a large public utility holding company. This is a type of corporate structure which has long been a matter of concern to the people of this country and to Congress itself because of its effectiveness as a device by which small groups of individuals, through the use of other people's money, are enabled to dominate large segments of our national economy without any corresponding responsibility to the public which is so vitally affected by their operations, or even to their stockholders whose proxies they use to solidify their positions of power. If to this concentration of economic power there is added the tremendous power of influencing public opinion which goes with the operation of major broadcasting facilities, domestic and international, the result is the creation of a repository of power able to challenge the sovereignty of government itself. (Crosley, 1945, p. 34)

Although the transfer was approved, the Commission adopted new procedures to give it an opportunity to compare potential transferees in the future.

The Avco rule imposed a 60-day waiting period for any proposed transfer. During that time application from other organizations could be filed which would offer to pay the same purchase price. The F.C.C. then would hold hearings "to determine which of the applicants is best qualified to continue the operation of the licensee" (Crosley, 1945, p. 26).

The commission optimistically predicted that the procedure would have only beneficial consequences.

> It is believed that such a procedure will more adequately enable the Commission to carry out the congressional intent that the best qualified person be licensed for each available frequency. No harm will result to the transferor since he either sells his station to the person he selected under the terms and conditions specified in his contract, or he sells to another person on the same terms. . . . The only change which this procedure will cause is giving the Commission some real voice in the selection of a proper successor. (Crosley, 1945, p. 27).

The Commission failed to recognize that this procedure would impose a substantial delay in the transfer of broadcast properties. In addition to the 60-day waiting period, substantial time would be required for the hearings to determine the best qualified applicant. This could be inconvenient or even disastrous for sellers, who needed access to the cash which could be realized through the sale of a station. It also could be discouraging to buyers, who would not wish to have such a substantial potential obligation limiting their options for other investments until the F.C.C. finally reached a decision. Although the Avco rule might have had benefits for the public by encouraging transfers to better qualified licensees, the industry universally viewed it as cumbersome and obstructive. Within a few years the Avco rule was abandoned, and the F.C.C. never again attempted to control transfers.

Repeal of the Avco Rule—Letting the Marketplace Uphold the Values

The opposition to the Avco rule was immediate. The primary argument used against the rule was that it failed to serve the intended value of providing superior service to the public. The opponents stressed that the Avco proceedings actually resulted in inferior service. Within 2 years the Federal Communications Bar Association sent a formal letter to the F.C.C. calling for the repeal of the rule.

> During the past two years that this system [the Avco rule] has been in effect, relatively few competitive applications have been filed. Accordingly, it appears that the original purpose of the Avco procedure has not been fulfilled. Moreover, where the transferor does not wish to sell to the competitive applicant, despite the Commission's decision favoring it, it appears that the Avco procedure and the

consequent delay during the initial waiting period of sixty days subsequent to the filing of the application, broadcast licensees encounter considerable difficulty from a business standpoint. Station personnel are left in a state of uncertainty regarding their ultimate employment by the prospective purchaser and advertisers are similarly uncertain regarding the continuance of the management with whom they have been dealing. (Federal Communications Bar Association, 1948, p. 159)

The industry continued to gather information on the cases where the rule appeared to have an impact which was contrary to the public interest.

Examples of the rule failing to serve the public were gathered and were recounted at every opportunity. The long-delayed and ultimately aborted transfer of KMED is an example of a negative situation the rule created.

Mrs. W. J. Virgin "sold" KMED to Luther E. Gibson, owner of KHUB Watsonville and KSLI(FM) Salinas, Calif., for $250,000 plus. A group of Medford businessmen filed a competing bid and won F.C.C. approval. But Mrs. Virgin refused to sell, claiming the Medford group's offer was not on the "same terms and conditions" as Mr. Gibson's.

The Medford group ultimately withdraw and F.C.C. granted reinstatement for the original Virgin-Gibson application—on the condition that it go through the advertising procedure again. Meanwhile their contract expired and they were unable to come to new terms. Mr. Gibson told F.C.C. that all the 2½ years of proceedings had brought him was an out-of-pocket expense of some $40,000. (Crater, 1949, p. 24)

The opposition movement was successful. In 1949, a little less than 4 years after its adoption, the F.C.C. announced it no longer would hold hearings to select the best qualified transferee. Henceforth, transfers would be disapproved only if the transferee failed to meet the Commission's minimum standards.

The broadcast industry was happy with its victory, but it wanted to insure that the F.C.C. never again tried to take any action to impede license transfers arranged by broadcasters. In 1952 Congress enacted the first package of amendments to the Communications Act of 1934. Included in these amendments was an addition to Section 310.

In acting [on any transfer application] the Commission may not consider whether the public interest, convenience and necessity might be served by the transfer, assignment, or disposal of the permit or license to a person other than the proposed transferee or assignee. (47 U.S.C. Section 310(d))

Now, even if some future F.C.C. wanted to, it could not establish a new Avco Rule.

The broadcasters were very happy that Congress issued this directive to the F.C.C. The legislation closed the discussion of comparative evaluation of transferees. Now the F.C.C. was prohibited from resurrecting an AVCO-type process for making value judgments on the relative qualifications of potential transferees. The Commission's next major attempt to safeguard against rampant speculation in broadcast licenses was the 3-Year Rule.

The 3-Year Rule—Should Long-Term Commitment Be Valued?

The 3-year rule was adopted in 1962. It required that hearings be held on any proposed broadcast license transfers if the license had been held for less than 3 years. Although the rule did not prohibit more rapid transfers, the threat of hearings (and their accompanying uncertainty and delay) was enough to eliminate virtually all short-term transfers.

The rule sought to address two concerns: that the rapid turnovers of stations might be reducing program quality, and that the increasing number of turnovers during the late 1950s reflected undesirable "trafficking" in licenses by speculators whose interest in quick profits would prevent them from providing substantial program service.

The rule was in effect from 1962 until early 1982. As Figures 16.1 and 16.2 indicate, it did have the effect of leveling the rate of growth in the number of radio and television station transfers. Outside the period when the rule was in effect, the rate of growth in the number of transfers was much higher.

The broadcast industry was consistent in its objections to the rule. The comments ABC filed with the F.C.C. give a good representation of the bases for this opposition.

> The three year rule [is] a good example of a regulation that has outlived its usefulness [and that the rule now stands as] a needless inhibition on normal business and marketplace forces in the radio and television industries. (Amendment of Section 73.3597, 1982, p. 1082)

In repealing the rule, the F.C.C. went beyond asserting that it was unnecessary to argue that the rule was really counterproductive.

The view that the rule was harmful was most clearly expressed by F.C.C. Chair Mark Fowler.

> The purpose of the rule may have been laudable at the time it was developed by the Commission. But its results proved quite the opposite. It punished rather than rewarded entrepreneurs that tried to improve the service (and the profitability) of one station and move along to another station. It thereby discouraged entry and exit in the broadcasting business. The Three-Year Rule really became the seven-minute penalty box for the broadcast industry. Instead of removing a contentious player from the game, it probably removed the best. Today's action shows you can bring regulation in line with what really goes on in the marketplace and best serve the public's interest. (Fowler, 1982, p. 1091)

In the deregulation of the 1980s the investor who is able to reap a rapid profit from the transfer of a broadcast property is asserted to be demonstrating unusual ability to serve the public. The behavior which was viewed as undesirable (and perhaps even reprehensible) now is seen as exemplary. Rather than an evil to be avoided, "trafficking" is now seen as the mark of "the best."

This provides an interesting resolution to the value dilemma proposed above: Should policy promote the financial profit of the broadcaster, or should it promote the "public interest, convenience, and necessity." Fowler asserts that, with regard to license transfers, there is no dilemma. The financial profit of the broadcaster is asserted to come only from the broadcaster's unusual skill in serving the public.

The pace of transfers has accelerated with the termination of the F.C.C.'s "3-year rule" and opened new opportunities for rapid station trading. Not everyone agrees with Fowler's expectation that this will reward the best. Del Regno observes

> Removing the three-year rule has made it easier for adventurous souls to come in there and to know that they can get out of this business and sell off the company, or parts of it, in the next year or two. Their investment is protected; perhaps more than protected. Perhaps they can make money on acquiring a company and selling off its parts, and they can do that in an efficient manner, a speedy manner. They don't have to wait it out. ("Money on Their Minds," 1986, p. 79)

This has had an impact on access to the industry. No longer is the transfer of a station limited to potential licensees who, of necessity, must have a commitment to the long-term operation of the station. Now stations can be sold to investors who are looking for short-term profits from either the operation or the early sale of the station, rather than from long-term gains. It is likely that such short-term and profited-oriented investors will not have a strong commitment to the long-range values which were emphasized in the early days of broadcasting.

Policies on Diversity of Control

Underlying many of the F.C.C.'s actions is a generally shared value which holds that the public should have access to diverse sources of information. In broadcasting, one of the ways to encourage this diversity is to establish policies which encourage licenses to be held by people from many different groups. In recent years, some of the policies established to achieve this goal have been challenged and modified. The primary justification for the changes has been that diversity of control in broadcasting is less relevant today, with the wide range of media available to the public, than it may have been when these limitations were established, with radio, television, and newspapers being the only media choices.

Local limitation on media control. Many of the limits on control have been established to insure that the public in each community has access to media controlled by different organizations. The primary rules to implement this goal are

> *The Duopoly Rule:* No licensee can have two stations in one service (i.e., two AM stations, two FM stations, or two TV stations) whose signals overlap.
>
> *The One-to-a-Market Rule:* No licensee can have more than one station (AM, FM, or TV) in a market. [Except that enterprises which had combinations of stations prior to the adoption of this rule may continue to operate those stations until they are sold.]
>
> *The Rule Prohibiting Cross-Ownership of Cable Television and Television Stations:* No licensee can have a television station whose signal overlaps with a cable television system it owns.
>
> *The Rule Prohibiting Cross-Ownership of Newspapers and Broadcast Stations:* No licensee can have a broadcast station (AM, FM, or TV) in a market where it owns a daily newspaper. [Except that most combinations established prior to the adoption of this rule may continue.]

There appears to be general agreement on value of promoting diversity within each market. The only controversy surrounding these rules is over whether they should be waived in individual situations because of the meritorious service of a particula owner.

National limitations on media control. Until 1984 the F.C.C. imposed strict limits on the number of stations that it would license to one organization. An organization was permitted a maximum of seven AM stations, seven FM stations, and seven TV stations. These rules were designed to discourage broadcast monopolies and to encourage local control of stations.

The numerical limits were raised to 12 stations in each service in 1984. After 1990, there will be no limit on the number of stations a single organization can control. In justifying its decision to remove this limit, the F.C.C. points to the growing diversity of media outlets in addition to radio and television. It also places great weight on what it sees as the quality of the programming of the group station owners. Commission Chair Fowler emphasizes this in his supporting statement.

> It is commonplace in Washington to remind one another that bigness is not necessarily badness. Today we are bringing that maxim to the broadcast ownership context. Bigness is not necessarily badness, sometimes it is goodness, sometimes it is just bigness and nothing more. (F.C.C., 1984, 58)

In keeping with the deregulatory mood of the 1980s, neither Fowler nor the Commission entertained the possibility that bigness *might* be badness sometimes.

Raising the limit to 12 stations has encouraged a trend where station transfers tend to result in their being acquired by new owners who control substantially more stations than were controlled by the previous owners (Foley, 1987a). When the limits are removed entirely, this trend should accelerate. Whatever values there may have been to encouraging local ownership have been lost in the "bigness is not necessarily badness" movement.

Transfers to minority group members. Another value which is generally accepted is that diversity of control of licenses among racial and ethnic minorities is desirable. In recent years there has been extensive controversy over the extent to which the F.C.C. should (or may) adopt policies which would implement this value by encouraging the ownership of stations by women or by minority group members. Citizen groups have criticized the F.C.C. for failing to take appropriate action. Broadcasters have seen the F.C.C.'s action as undesirable tampering with the marketplace. In the area of license transfers, there are three policies which favor transfers to minority group members.

Preferences for minority group members and women. In comparative hearings, the F.C.C. has given some preferential weight to applications from minority group members and women. Recently (Steele v. F.C.C., 1985), the Court of Appeals has ruled that the F.C.C. has exceeded its statutory authority in allowing preferences for applications from women. Wilson provides an extensive review of these preferences and concludes, "minority and gender preferences are certainly within the bounds of the public interest standard" (Wilson, 1988, p. 114).

Distress sales. Ordinarily, the F.C.C. will not allow a station to be transferred while the F.C.C. is conducting hearings on allegations that the station has violated the F.C.C.'s rules. Transfers are allowed for such stations if the license is transferred to a new licensee controlled by a member of a minority group, and if the transfer is made for significantly less than the appraised value of the station.

Tax preferences. Firms which transfer licenses to enterprises controlled by minority group members can obtain some tax benefits. If the F.C.C. certifies the transfer, the seller can be granted a waiver of any capital gains tax which would have been due on the transaction if the seller invests in another broadcast station within the next 2 years. Since station values have increased dramatically in the past decade, the capital gain realized from transfers can be substantial. This tax preference can amount to a saving of millions or tens of millions of dollars.

Mergers and acquisitions. In the past decade, the United States has seen many corporate merges and takeovers. Many times these have been hostile takeovers which were accomplished despite the objections of the board of directors of the firm being acquired. For many years firms with broadcast licenses were considered poor targets for hostile takeovers because of the requirement that the F.C.C. must approve any transfer of their licenses.

In keeping with its campaign for deregulation, the F.C.C. has worked to simplify its procedures on such takeovers. "The Commission's chief concern about the hostile takeover phenomenon was that its procedures would inhibit the free operation of the market for control of corporate broadcasting licenses" (McGill, 1988, p. 41). The F.C.C. has adopted policies to expedite the transfer of licenses in takeovers (friendly or hostile), so long as the new licensee meets its basic standards (F.C.C., 1986). These policies are designed to insure that take-overs are not delayed by significant consideration of whether the associated license transfers would serve the public interest.

VALUE ISSUES

Rights in Conflict

The root of the value issues which underlie the debates over the policies outlined above is found in the First Amendment to the U.S. Constitution. "Congress shall make no law . . . abridging the freedom of speech, or of the press. . . ." The absolute prohibition of this provision worked well to protect rights with the communication technologies which existed prior to broadcasting. With the coming of radio, it raised a serious dilemma: Was it designed to protect the rights of the listener or the rights of the broadcaster? In most situations the rights of these two groups are diametrically opposed. Any rights of expression given to the listener come at the expense of the broadcaster's right to control the content of programming. It is impossible for both listeners and broadcasters to have maximum right to free speech and free press.

The resolution of this dilemma in the Communication Act of 1934 (and in its predecessor Radio Act of 1927) was to require that broadcast licensees operate "in the public interest, convenience, and necessity." In exchange for operating their stations within these guidelines, they were allowed substantial freedom to select the programs they would air, subject to the controls which might be exerted through the ever-present (albeit remote) threat that their licenses might not be renewed.

Over the years the view has evolved that the rights of the listeners are best served by creating an environment in which they can obtain diverse programming. They are not allowed direct access to the airwaves. Rather they must depend on the broadcast licensees to provide programming for them.

The result of this is that the listeners can only be served if the broadcaster decide to do so. There are three alternative general policy strategies available for achieving this balancing of rights: (a) broadcast licensees could be selected by their deep social commitment to providing service; (b) broadcasting could be regulated so that all licenses were forced to provide substantial service, regardless of their natural inclinations; or (c) the competitive market for programming could function in such a way that it encourage broadcasters to provide service.

The first alternative basically would require that broadcast licensees would be willing to sacrifice their own rights to control the content of the media in order to promote the rights of their listeners. It has been practiced to varying degrees over the years in the selection of licensees for new frequencies, where the selection is based on a comparison of the merits of the various applicants. Many broadcasters and broadcast corporations have shown great willingness to try to serve the needs of their audiences. As indicated above, these comparative criteria are not used when a license is transferred by an existing licensee.

The second alternative was used as the primary means of control prior to the wave of deregulation that began in the 1970s. It is based on the concept that broadcasters (or at least some broadcasters) must be forced to recognize the rights of the public as well as their own rights. The primary means for this regulation was extensive hearings into licensee conduct and the threat of non-renewal of licenses.

The third alternative has grown in favor with the coming of deregulation. Its proponents assert that the marketplace provides better control than the regulators could provide. Basically, it asserts that the apparent conflict between the rights of the listeners and the rights of the broadcasters is illusory. It maintains that the broadcaster's self-interest in maximizing its profits will result in the programming which best serves the needs of the listeners.

The discussion of the extent to which the F.C.C. should be involved in regulating the transfer of stations addresses these fundamental value concerns in the control of broadcasting.

CONCLUSIONS

Given the above analysis it is clear that both the substantial profits which can be made from the sale of broadcast stations, and the transfer process itself, encourage the trend to think of broadcast stations primarily as revenue-generating investments, rather than as instruments of democratic communication whose primary goal should be to serve the public interest, convenience, and necessity. The astronomic prices paid for entry into broadcasting insure that financial analysts will have a major role in any decision about acquiring a station. There are no countervailing factors to give significant weight to the nonfinancial aspects of the transaction.

Ultimately, it is not just the station that is being sold as a commodity; it is the audience as well. The payment is for the potential of reaching the eyes and minds of the public: not for reaching them with messages relevant for the great political and social issues of the day, but for reaching them with the commercial messages of the sponsors who wish to advertise. The recent National Association of Broadcaster's guide to purchasing a broadcast station, subtitled "Everything you Want—and Need—to Know but Didn't Know Who to Ask," shows the extent to which this pattern of thinking pervades the industry. The guide provides detailed information on the financial aspects of broadcasting, with no mention of any public interest obligations.

> Although broadcasting is very demanding in the combination of management and people skills required and may be more sensitive to general trends in the economy than many industries, there are many reasons why acquiring a broadcast station can turn out to be a good investment. . . .
>
> The balance for which a successful broadcaster strives is to air programming that will attract the highest number of listeners or viewers so that the station can charge advertisers the highest price for commercials. (Krasnow, Bentley, & Martin, 1988, pp. 1, 3)

In the spirit of the deregulation of the 1980s, this guide provides detailed information on the financial aspects of broadcasting, with no mention of any public interest obligations.

For both the regulator and the broadcasters the values have shifted. From an original concern with balancing the broadcaster's need to make a profit with the obligation to serve the public in exchange for the exclusive use of a frequency, we have moved to the assumption that operating the station in ways which maximize the price for which it could be sold will result in programming which best serves the public interest. This assumption is being widely implemented without gathering evidence to demonstrate that it is valid. The historic concern for the social values which could be served by broadcasting is being replaced by the current emphasis on economic values. In the process the social values are in danger of being lost.

REFERENCES

Crater, R. (1949). Avco Repeal. *Broadcasting, 36,* 23–24.

Crosley and the Aviation Corp. (1945). 11 F.C.C. 3.

Federal Communications Bar Association. (1948). Letter to the F.C.C., 15 October, as quoted in Federal Communications Commission, 1959, 159.

Federal Communications Commission. (1982). Amendment of Section 73.3597 of the Commission's Rules (Applications for Voluntary Assignments or Transfers of Control). 52 RR 2d 1081.

Federal Communications Commission. (1984). Amendment of Section 73.3555 of the Commission's Rules Relating to Multiple Ownership of AM, FM, and Television Broadcast Stations, 100 F.C.C. 2d 17.

Federal Communications Commission. (1986). Tender Offers and Proxy Contests, 59 R.R. 2d 1536.

Foley, J. M. (1987). "Selling the Store: Policy Implications of the 1986 Bonanza in

Television Station Transfers," Telecommunications Policy Research Conference, Airlie House, Virginia.

Fowler, M. S. (1982). Concurring statement in Amendment of Section 73.3597 of the Commission's Rules 52 RR 2d 1081, 1091.

Krasnow, E. G., Bentley, J. G., & Martin, R. B. (1988). *Buying or building a broadcast station: Everything you want—and need—to know, but didn't know who to ask* (2nd ed.). Washington, DC: National Association of Broadcasters.

McGill, I. G. (1988). The market for corporate control in the broadcasting industry. *Federal Communications Law Journal, 40*, 39–87.

Money on Their Minds. (1986, April 28). *Broadcasting*, pp. 74–81.

Proceedings of the Fourth National Radio Conference and Recommendations for Regulation of Radio. (1925). As quoted in F.C.C., 1959, 22.

Smith, B. B. (1945). The radio boom and the public interest. *Harpers Magazine, 191*, 315–320.

Steele v. F.C.C. (1985). 770 F.2d 1192 (D.C. Cir.).

Wilson, L. C. (1988). Minority and gender enhancements: A necessary and valid means to achieve diversity in the broadcast marketplace. *Federal Communications Law Journal, 40* (1), 89–114.

chapter 17
Latent Policy and the Federal Communications Commission

Sven B. Lundstedt
Ameritech Research Professor
Professor of Public Policy and
    Management, and
    International Business
School of Public Policy and
    Management and Faculty
    of Human Resources and Management
The Ohio State University

Michael W. Spicer
Dean and Professor
Department of Public Administration
College of Urban Affairs
Cleveland State University

It may come as no surprise to learn that some policies and values in complex organizations are advertently or inadvertently covert. Often they are simply the result of informal decision making and the consequence of a form of decision-making shorthand resulting in a failure to make them manifest for quite legitimate reasons. At other times, they may be deliberate covert strategies for good or ill. In any event, we think they are an important category of decision rules which need to be recognized in the study of organizations, public or private. *Latent policy* (Lundstedt, 1976) means an intentionally or unintentionally concealed policy of action of which an individual may be aware or unaware. As a matter of historical background, the basic concept of latent process has a long record of use in psychology and sociology in the analysis of personality, social processes, and social systems. In this sense, latent policy is a type of social function. As Merton explains, "The basic requirement is that the object of functional analysis represents a standardized item, such as social roles, institutional processes, cultural patterns, culturally-patterned emotions, social norms, group-organization, social structure, devices for social control, etc." (Merton, 1949). This particular functional definition would view policy as a kind of social or organizational norm as part of an institutional process. Latent functions, including values, are among those important observed consequences of social behavior which influence the adaptations or adjustments which occur in social systems. This definition allows the term latent policy to be used in a manner which gives it a wider theoretical justification and basis. The words *function* and *policy* are interchangeable.

Other characteristics of latent policy are also important. First, some seemingly irrational behavior in organizations can be explained by the distinction between *manifest* and *latent* policy. Individuals may not always act irrationally, even if they appear to do so; they may actually be guided by different kinds of rational policies and values which they have decided cannot be expressed in normal ways. The distinction between manifest and latent also directs attention toward the unanticipated consequences of manifest policies which lie beyond, and behind, the usual formal procedures in organizations.

Manifest and latent policies may be further broken down into convergent and divergent forms (Lundstedt, 1976). By comparing them to a given focal policy, divergent policies are in conflict and convergent ones in harmony with that focal policy. Manifest convergence and divergence among policies commonly reflect open communication in the administration of policy. The values of free and open exchange of information often underlie, even encourage, manifest policy divergence in public debates about policy alternatives. This familiar process is made exceedingly more complex if the added dimension of a latent policy is introduced, especially one which may diverge markedly from a given manifest focal policy.

Latent convergence is simply tacit agreement between a policy and its dominant focal policy system. In this case the values reflected in a given policy are said to be congruent with existing dominant values. Latent divergence (tacit disagreement), however, arises from different attitudes and behavior, and it has a different potential effect upon efforts to form and implement policy. For example, if confidence is lost in a given manifest policy system because it appears to fail to operate effectively for any reason, the probability that a latent divergent response will appear to compensate for it may increase. With it may also come an increase in mistrust, poor communication, misunderstanding, hostility, and policy conflict, because key actors have gone "underground." Latent divergent responses in a free society can also become covert behaviors which may lead to counterproductive strategies which are indiscriminantly and even dangerously covert. Watergate and similar events are interesting recent examples of this kind of situation. In the absence of evenly administered policies which strengthen democratic values and procedures, this outcome is always a greater natural risk.

Of particular interest are the factors which underly the emergence of latent policy and latent values in an organization. In order to shed light on these factors, it may be useful to examine a concrete example of latent policy in the public sector. In particular, we wish to examine the policy of the United States Federal Communication (F.C.C.) with respect to local ownership of television stations.

F.C.C. AND LOCAL OWNERSHIP

The manifest commitment of the F.C.C. to local ownership can be traced back to early regulation of radio. It stems from a concern over the dangers of monopoly and a widely held belief that broadcasting needs are best served when broad-casting stations are owned by local residents. Local ownership of television stations constitutes a major "doctrine" of F.C.C. policy (F.C.C. 1957). The F.C.C. has argued that full-time participation by owners in station operations is likely to lead to a "greater sensitivity to an area's changing needs" and to "programming designed to serve those needs." In addition the F.C.C. argues that station ownership by those who are local residents may still be of some value even when there is no substantial participation in operations. Thus it would appear to be the manifest policy of the F.C.C. to support local ownership of television stations.

In the past two decades, however, despite this manifest policy, locally owned television stations have become "curios rather than the industrial norm" (Bunce, 1976). The percentage of television stations locally owned has declined from 55.2% in 1956 to 33.2% in 1974 (see Tables 17.1 and 17.2). In the largest 50 markets, which account for 50% of television audiences, the percentage of local ownership has declined from 43.9% in 1956 to 27% in 1974 (see Tables 17.3 and 17.4).

Closer analysis of the data reveals that the decline in the percentage of local ownership from 1956 to 1966 is statistically significant both for all markets and the largest 50 markets. The decline from 1966 to 1974 is, however, not statis-tically significant for either. At first glance, this might indicate that the F.C.C. has succeeded in at least slowing the decline of local ownership.

However, available evidence suggests a different conclusion. For example, in 1964 and again in 1965, the F.C.C. proposed rule amendments to limit the sale of television stations to owners of groups of stations, but in both cases these proposals were withdrawn in the face of opposition from group owners. Richard Bunce argues that "local ownership is often proclaimed but almost nowhere enforced" (Bunce, 1976). If anything, the evidence suggests that the F.C.C. has played a role in aiding the decline of local ownership. A major function of the F.C.C. is the receiving of applications for new stations. If the F.C.C. were to act on its manifest policy on local station ownership, then it would follow that the F.C.C. would give preference to an owner-manager rather than an absentee owner. A statistical study of F.C.C. license awards from 1967–70 indicates that, other things being equal, an applicant that listed local residents as owners was 25% less likely to be granted a license than a competing applicant with no local

Table 17.1.   Local and Nonlocal Ownership of Television Stations in All Markets

Location of ownership	Percentage in 1956	Percentage in 1966	Percentage in 1974
Locally owned	55.2	33.2	33.2
Nonlocally owned	44.8	66.8	67.8

From *Television in the corporate interest*, by R. Bunce, 1976, New York: Praeger. Copy-right © 1976. Reprinted by permission.

Table 17.2.   Percentage of Change in Local and Nonlocal Ownership of
Television Stations in All Markets

Location of ownership	Percentage of change 1956–66	Percentage of change 1966–74	Percentage of change 1956–74
Percentage locally owned	−22.0ᵃ	0	−22.0ᵃ
Percentage nonlocally owned	+22.0ᵃ	+1.0	+23.0ᵃ

ᵃstatistically significant ($p < .01$)

representation (Noll, Peck, & McGowan, 1973). Furthermore, promises of local programming also seem to reduce an applicant's chance of receiving a license—an interesting finding in light of the F.C.C.'s rational for local station ownership. The licensing behavior of the F.C.C. seems clearly inconsistent with its manifest policy.

A further inconsistency appears when one examines F.C.C. policy for renewing existing licenses for television stations. The one real sanction the F.C.C. has in pursuing policy objectives, such as local station ownership and local programming, is refusal to renew a license. However, the F.C.C. has proved generally to be very reluctant to use this sanction, and license renewal is virtually automatic. Indeed, in 1970 the F.C.C. issued a policy statement making it clear that preference in license applicants would be given to incumbents. This policy statement was later found to be illegal by the District of Columbia Court of Appeals.

In summary, while it is the manifest policy of the F.C.C. to support local ownership of television stations, its actions are not supportive of local ownership and in fact appear to favor absentee ownership. Why does this divergence between manifest and latent policy exist?

Table 17.3.   Local and Nonlocal Ownership of Television Stations in Largest 50 Markets

Location of ownership	Percentage in 1956	Percentage in 1966	Percentage in 1974
Locally owned	43.9	30.6	27.0
Nonlocally owned	56.1	69.4	73.0

From Television in the corporate interest, by R. Bunce, 1976, New York: Praeger. Copyright © 1976. Reprinted by permission.

Table 17.4.   Percentage of Change in Local and Nonlocal Ownership of
Television Stations in Largest 50 Markets

Location of ownership	Percentage of change 1956–66	Percentage of change 1966–74	Percentage of change 1956–74
Locally owned	−13.3[a]	−3.6	−16.9[a]
Nonlocally owned	+13.3[a]	+3.6	+16.9[a]

[a]statistically significant ($p < .01$)

FACTORS UNDERLYING LATENT POLICY

One important factor is vagueness in statutory guidelines for F.C.C. policy. The Communications Act of 1934, which established the Commission, provides only that the F.C.C. act in the "public interest, convenience, and necessity." The vagueness of the Congressional directive allows the F.C.C. the discretion to evolve its own policies and hence make possible a divergence between manifest and latent policy. Also important is the relative lack of visibility of F.C.C. decision making. While hearings are public, F.C.C. decisions are unlikely to command much attention unless losers in the application process, or interest groups, are prepared to appeal the decision or put direct pressure on Congress. Furthermore, since there are a number of criteria by which license applications are judged, it is difficult to discern on a case-by-case basis what policy is being followed with respect to local ownership of television stations.

The vagueness of enabling legislation, and relative lack of visibility of F.C.C. decision making, create, we believe, the conditions necessary for the existence of divergent latent policy different from stated manifest policy. Whether these factors are sufficient to explain divergent latent policy, however, seems doubtful. It seems to us that such an explanation lies with an understanding of the political economy of the regulatory process.

Two competing hypotheses are commonly put forth concerning the behavior of regulatory agencies (Posner, 1974). First, the agency is seen as pursuing its conception of the public interest, albeit sometimes incompetently. Second, the agency is seen as maximizing the welfare of the groups it regulates. However, as Porter and Sagansky (1976) note, to assume that a regulatory agency's decisions accurately reflect its goals is to ignore the constraints on its decision-making process. Such constraints are important as a potential source of conflict between formally stated goals and ongoing behavior within an organization.

A particularly important constraint on regulatory decision making is the supply of information available. Whatever the goals of the F.C.C., its activities require large amounts of information. Information is, however, a scarce and costly resource. Given the complex and often technical nature of the task of regulating television, and the large number of firms involved, it seems reasonable to hypothesize that officials in the executive branch, including the F.C.C. itself, should seek to economize on the information obtained to meet F.C.C. decision-making responsibilities.

One way of doing this is to appoint officials to the F.C.C. who have had prior experience in broadcasting, and who are themselves sources of information. An examination of the career backgrounds of F.C.C. officials lends credence to this idea. In a recent statement to the Congressional Subcommittee on Oversight and Investigations (1975), James Rieger of the General Accounting Office presented data indicating that some 50% of commissioners and administrators appointed to the F.C.C. between 1960 and 1975 had prior direct or indirect employment in the communication industry. The term *indirect employment* means employment in enterprises which perform services for the industry such as law firms or consulting companies. A recent survey undertaken by Common Cause (1976) indicates that, during fiscal years 1971–1975, five of the seven commissioners named to the F.C.C. came from the communications industry, and that, in 1975, 51% of F.C.C. employees in GS 15 or higher positions used to work in the industry (see Table 17.5).

Another way of economizing on information costs is for F.C.C. officials themselves to collect information from cheaper sources. The cost of information from any supplier is in part related to the readiness of that supplier to make available such information, and in part to transactions costs. The communications industry, on this basis, clearly has a competitive advantage. First, it is willing freely to supply certain types of information, since the supply of information gives the industry at least some influence on F.C.C. policy. Second, in light of the industry background of many F.C.C. officials, it follows that these officials may be able to secure information from industry relatively more easily than from other sources. It is, therefore, not surprising to find that, according to Krasnow and Longley (1973), the Commission depends on industry trade associations and broadcast licenses for much of its information. This information is transmitted by consultative groups, publication of views in the trade press, liaison committees of the Federal Communication Bar Association, social contacts and visits to

Table 17.5.   Career Background of F.C.C. Officials

Commissioners		Senior Employees	
Total Number Appointed	Number from Industry	Total Number Surveyed	Number from Industry
7	5	37	19

From *Serving two masters,* by Common Cause, 1976. Copyright © 1976. Reprinted by permission.

offices of the Commissioners, informal discussion at state broadcaster and trade association meetings, and the formal submission of pleadings and oral argument. Crucial in mediating the information flow is the nature of the commission–industry relationship. Contacts with the general public through citizens groups and court orders seem more likely to be viewed as threatening than the relationship between the commission and the industry. According to Landis, direct contacts with industry are necessarily often productive of intelligent ideas, "whereas contacts with the general public are rare and generally unproductive of anything except complaint" (Landis, 1960).

Information costs are thus reduced by a combination of recruiting a significant proportion of F.C.C. officials from the communications industry and a heavy reliance on the industry as a cheap supplier of information. Such economics, however, have important psychological consequences for policy. Information will usually elicit both cognitive and affective responses to it and both play an important role in shaping an individual's perceptions and preferences. Given the way in which F.C.C. officials acquire information, it seems reasonable to hypothesize that the past and present interaction between F.C.C. and industry officials leads to a latent convergence of their perceptions and preferences concerning F.C.C. policy. Indeed, former F.C.C. commissioner Nicholas Johnson has charged that there exists a "subgovernment" of industry lobbyists, speciality lawyers, trade associations, trade press, congressional subcommittee staff members, and commission personnel who dominate F.C.C. policy.

Some evidence of latent convergence of F.C.C. and industry perceptions and policy preferences is suggested by the career patterns of commissioners after they leave the F.C.C. Lichty (1961) found that most commissioners eventually entered broadcasting or related fields after leaving the F.C.C. Recent studies by Noll et al. (1973) and Williams (1976) indicated a continuation of this trend. Also, Common Cause (1976) reported that four of the seven commissioners who left during fiscal years 1971–1975 took jobs in the communications industry.

Of course, all this does not mean that F.C.C. officials are in open collusion or corrupt. The evidence is simply indicative of a shared viewpoint, which is often an indirect outcome of our interest group focus in government.

Given the present structure of the television broadcasting industry, expressed industry views on local ownership are likely to be in large part those of multiple station owners. The shared viewpoint of F.C.C. and industry officials is, therefore, unlikely to be strongly supportive of local ownership and thus is likely to conflict with F.C.C. manifest policy on local ownership. This conflict between the perceptions and preferences of F.C.C. officials and manifest convergent policy provides the essential basis for divergent latent policy.

The emergence of latent policy is somewhat analogous to the emergence of an informal organization within formal organizational structures. Informal interaction usually stems from a failure of formal structures to handle interdependencies within or between organizations and may lead to the development of norms either supportive of, or in conflict with, those of the formal organization. As early as 1933, Elton Mayo observed how groups of workers may informally set rules concerning the pace of work, with little regard to formal incentive schemes. A variety of studies, too numerous to cite here, sustain these findings. These rules can be thought of as a form of latent policy.

In summary then, the F.C.C.'s demand for information leads to an economizing of information costs which in turn affects the perceptions and preferences of F.C.C. officials. The information supplied directly by the industry, or indirectly through F.C.C. officials formerly employed by the industry, tends to lead to a latent convergence of F.C.C. and industry viewpoints and thus to a conflict with existing manifest public policy. Latent policy is seen here as a means for F.C.C. officials to help resolve that conflict. This does not imply that F.C.C. officials act deliberately to thwart the public interest. It may simply be that F.C.C. manifest policy on local ownership is no longer seen by its officials as crucial to enhancing what they perceive as the public interest. If this is the case, then the basis for the survival of manifest policy on local ownership is far from clear. Its survival may stem in part from that natural inertia in the policy process which always seems to be present, resulting in a failure of formal structure to reflect ongoing informal processes which are ahead of it. It may be that manifest policy even serves a symbolic purpose in helping to support the legitimacy of government regulation of the television industry. The rationale for regulation must after all be in making the regulated do something that they would otherwise not do.

CONCLUSION

The foregoing analysis indicates in our view that the concept of "latent policy" is useful in policy analysis. Using the F.C.C. as a case study, we have suggested a number of factors that may lead to the divergence of latent policy from manifest policy. In the case of F.C.C. policy on local ownership of television stations, we have suggested that such divergence stems from vagueness in legislative guidelines, lack of visibility of F.C.C. decision making and a conflict between manifest policy and the perceptions and preferences of F.C.C. officials. A key factor in creating such conflict may be the nature of information used by the F.C.C.

The extent to which these types of factors explain divergent latent policy in other policy areas is not known, perhaps because of a preoccupation with manifest policy. It is our belief, however, that many policy actions which superficially seem to indicate incompetence or irrationality may in fact reflect latent policy at work in highly competent and rational ways. If this is the case, then effective policy analysis should focus not only on manifest but also latent policy in its many forms.

REFERENCES

Bunce, R. (1976). *Television in the corporate interest* (chap. 2). New York: Praeger.

Common Cause. (1976). *Serving two masters*. Washington, DC: Author.

Federal Communications Commission. (1965). *Policy statement on comparative broadcast hearings*, FCC 2d 394, July 28.

Krasnow, E. C., & Longley, L. (1973). *The politics of broadcast regulation*. New York: St. Martin's Press.

Landis, J. M. (1960). *Report on regulatory agencies to the president-elect*. Published as a committee print by the subcommittee on Administrative Practice and Procedure of the Senate Committee on the Judiciary, 86th Congress 2d Session.

Lichty, L. W. (1961). Members of the Federal Radio Commission and the Federal Communications Commission, 1927–1971. *Journal of Broadcasting, 6*, 23–34.

Lundstedt, S. (1976). *Latent policy*. Working Paper Series, College of Administrative Science, The Ohio State University.

Merton, R. K. (1949). *On theoretical sociology* (chap. 3). New York: The Free Press.

Noll, R. G., Peck, M. J., & McGowan, J. J. (1973). *Economic aspects of television regulation* (chap. 4). Washington, DC: Brookings Institution.

Porter, M. E., & Sagansky, J. F. (1976). Information, politics and economic analysis. *Public Policy, 24*, 203–307.

Posner, R. A. (1974). Theories of economic regulation. *The Bell Journal of Economics, 5*, 335–358.

Subcommittee on Oversight and Investigations of the House Committee on Interstate and Foreign Commerce and the Senate Committee on Commerce and Government Operations. (1975). *Regulatory Reform, 1, Quality of Regulators*, 94th Congress, 1st Session, November 6 and 7.

Williams, W., Jr. (1976). Impact of commissioner background on F.C.C. decisions: 1962–1975. *Journal of Broadcasting, 20*(2), 239–260.

Author Index

Subject Index